D0152368

The ecological century

The author at mid-century.

The ecological century

A personal appraisal

E. Barton Worthington

Clarendon Press Oxford
1983

Oxford University Press, Walton Street, Oxford OX2 6DP
London Glasgow New York Toronto
Delhi Bombay Calcutta Madras Karachi
Kuala Lumpur Singapore Hong Kong Tokyo
Nairobi Dar es Salaam Cape Town
Melbourne Wellington
and associate companies in
Beirut Berlin Ibadan Mexico City
OXFORD is a trademark for Oxford University Press

Published in the United States by
Oxford University Press, New York

British Library Cataloguing in Publication Data
Worthington, E. Barton
The ecological century.
1. Ecology
I. Title
574.5 QH541
ISBN 0-19-854556-8

Printed in Great Britain by
The Thetford Press Limited, Thetford, Norfolk

To Stella (Johnson) (1905–1978)

and

To Harriett (Stockton) (1925–   )

Without whose encouragement this book
would not have been written.

Preface

The eighteenth century was marked by enlightenment and the nineteenth by industry. The twentieth may go down in history as the ecological century, and hence the title of this book. What the twenty-first century may be labelled—progress or chaos—will depend on how well mankind learns to ensure a proper balance between himself and his environment.

The word 'oecology' appears to have been coined by Ernst Heinrich Haeckel in his *Natural history of creation* (1868), but it did not come into regular scientific usage until near the end of the nineteenth century when the letter 'O' (*Oikos* = house or habitation) was soon dropped after a while in Britain and America, but it is still retained by some continental writers.

From the time of Humboldt's *Kosmos* (1845–62) and Darwin's *Origin of species* (1859) a few botanists and zoologists who were concerned with the environment as well as with plants and animals were laying the foundations of ecology as a scientific discipline. For some time plant distribution and the reasons for it tended to make the running. O. Drud's *Handbuch der Planzengeographie* appeared in 1890, A. W. Schimper's *Plant geography on a physiological basis* in 1903, E. Warming's *Oecology of plants* (English edition) in 1909, and F. E. Clements *Plant succession* in 1916. In America the subject became related to the conservation movement which was already under way with the Yellowstone National Park opened in 1872. American ecologists such as Clowes and Schantz were also linking conservation with agriculture, thereby creating another facet of ecology.

By the 1920s ecology was coming into its own and providing a scientific cloak for what had long been known as natural history. In plant science there appeared in 1923 A. G. Tansley's *Practical plant ecology*, in 1926 E. J. Salisbury's *Geographical distribution of plants in relation to climatic factors* (in the *Geographical Journal*), and in 1927 E. J. Russell's *Soil conditions and plant growth.*

By contrast the early development of animal ecology appears to have been slow, but its germs had been growing steadily under other names. Notable was fishery science in its studies of fish populations in relation to their physical environment, their food supply, and predators. Economic and medical entomology, marine and freshwater biology also had strong ecological components. The publication in 1927 of Charles Elton's *Animal ecology* became a landmark. Meanwhile human ecology had been launched as a subject in 1922, though not under that title, by A. Carr-Saunders' *The population problem*.

From the 1930s onward ecology as the mutual relations between living organisms and their environment slowly and steadily gained the respect of conventional biologists, but it was little known to the lay public until the third quarter of the century, when the environmental revolution got into its stride. A main factor in this was the phenomenal growth of communications. Travel became popular; photography, radio, and television brought interest into nearly every home. Then the term 'ecology' came to mean all things to all men and women. However, there has often been a tendency for emotions rather than factual evidence to take over, and this has sometimes led to bad conservation in addition to denigrating the meaning of ecology. In the late 1970s we had the phenomenon of 'Ecology Parties' entering the political arena in several countries. Some ecologists today say 'It is all moving our way now'; but there is danger of losing sight of ecological realities.

It fell to my lot to witness most of the evolution of ecology during this century. I was educated during the years when the discipline was becoming recognized, then undertook ecological exploration, particularly in the tropics. During the middle years I participated in the thinking of what ecological concepts mean in terms of conservation of resources and human progress, and later applied the concepts in practical affairs. It seemed therefore worth writing down some of these experiences. They may even help future workers who aim to make our world a place fit for all people to live in and enjoy.

The subjects discussed may at first sight appear somewhat disparate, but there is a watery connection running through most of them: water for fisheries in the African lakes, water for science and recreation in the Lake District, the control of great rivers like the Nile, man-made lakes and what they mean for people as well as for the environment, water and disease, water for beauty. This comes from my original training and experience as an aquatic biologist, but as years passed the fascination of the aquatic environment spread to the land, and from plants and animals to the human species, and even to large engineering works.

The story revolves not only around ecology as a method of thought and as a practical science, but also around the latter days of colonial history. 'Development and welfare' had come to the fore, not only in British but in varying degrees also in French, Belgian, Dutch, Portugese, and Spanish colonies. This movement, starting immediately after, and in the case of United Kingdom during, the Second World War was followed, sooner than most of us expected or thought fit, by the winds of change. Independence of the colonies, on the way by the early 1960s, was to a large extent completed within 15 years. The transitional period, the circumstances leading up to it, and what followed afterwards is already becoming a fashionable topic among students of political history. It was closely observed by colonial administrators, but not many of them have committed their experiences to paper. There are even fewer records by scientists of the changes in the developing world. The processes of change are still accelerating, and not always in satisfactory directions: for example, with the exception of a few pelicans crying in the wilderness the current catch-phrase of 'technological transfer' implies trying to establish in poor countries the kind of facility to which we have become accustomed in rich countries. 'Appropriate technology' is still slow in its appeal.

But what is the true result of the ecological century? I think it is the coming together of the many subjects affecting mankind and his environment. During the age of rapid expansion of the sciences and of economy in the developed—some would say over-developed—world, the subjects tended to separate too widely. Now physics and geology are coming to terms with economics, chemistry with biology, engineering and agriculture with medicine and the social sciences. There is still far to go in this but as integration proceeds the advice available to those who have to take decisions for the future becomes progressively more balanced. Ecology in its broad sense has pushed forward the two essential processes, first of analysis, then of synthesis.

It must be emphasized that no chapter in this book purports to provide an up-to-date scientific account of any subject discussed. Each chapter relates the situation as it appeared during the few years which figure in the chapter's title. Every one of the subjects discussed is evolving and most are expanding as the few students of the past turn into the many of today.

The chronological sequence of the chapters has been adopted for two reasons: first it may have historical value in illustrating the overall development of ecology from slender beginnings to a powerful world movement, and second it had profound effects on my own thinking and work, building up each activity on what had gone before. I have attempted to avoid autobiography, but various friends on reading the draft suggested that a summary of the author's childhood and education ought to be included to indicate why he became a zoologist, then a hydrobiologist, then an ecologist. Under pressure I have complied in a prelude (Chapter 1).

When writing about developing countries, especially Africa, there is today a problem about geographical names since many of those which have become entrenched on maps and in literature have been replaced since independence, often in honour of modern presidents or personalities in place of past eminences. Some of the new names have a short run—like Lake Idi Amin Dada for Lake Albert. So in this book the names which were in current use at the time are used with a few exceptions, but with no disrespect for some of the new names which have replaced them.

By the same token persons are referred to by their names and titles at the time, although some changed later; and the photographs have been selected from those taken at the time.

It has been necessary to refer to quite a number of organizations, some with heavy titles. These are given in full when they first appear, but later reference is by initials, with a glossary on p. 6. References to literature have been kept to a minimum.

I am indebted to many people with whom contact was close during the periods covered by the different chapters and also to life-long friends who are concerned in these matters. Some of them are mentioned by name in the text. In addition a number of people have helped in the process of writing. I express grateful thanks especially to those who improved the drafts of particular chapters; they include Leonard Beadle (Chapter 2), David Le Cren (Chapter 4) Jack L. Harley (Chapter 6), and Max Nicholson (Chapter 9). R. S. Baxter kindly arranged for references in Chapter 11 to development projects undertaken by Sir William Halcrow and Partners to be checked. Rosemary Lowe-McConnell and my sister Elizabeth Motley have been through the whole text to its considerable advantage, and my former assistant in the IBP, Sue Darell-Brown, has kindly looked after the secretarial side.

E. B. W.

Sussex
October 1982

Contents

Illustrations

Abbreviations

AEF	Afrique Equatoriale Française
AOF	Afrique Occidentale Française
ARS	African Research Survey
CCTA	Commission de Cooperation Technique pour l'Afrique
CD&W	Colonial Development and Welfare
CHEC	Commonwealth Human Ecology Council
CSIR	Council for Scientific and Industrial Research
COWAR	Committee on Water Research
CSA	Conceil Scientifique pour l'Afrique
EAHC	East Africa High Commission
FAO	Food and Agriculture Organization
FBA	Freshwater Biological Association
IAHS	International Association of Hydrological Science
IBP	International Biological Programme
ICSU	International Council of Scientific Unions
IFAN	Institut Français l'Afrique Noire
IGY	International Geophysical Year
INEAC	Institut d'Elevage et d'Agriculture au Congo
IRSAC	Institut de Recherche Scientifique d'Afrique Centrale
ITE	Institute of Terrestrial Ecology
IUBS	International Union of Biological Sciences
IUGG	International Union for Geology and Geophysics
MAB	Man and Biosphere Programme
MIT	Massachusetts Institute of Technology
NCC	Nature Conservancy Council
NERC	Natural Environment Research Council
OAU	Organization for African Unity
ORSTOM	Office de Recherche Scientifique et Technique Outre Mer
SCIBP	Special Committee for the IBP
SCOPE	Scientific Committee for Problems of the Environment
UATI	Union of International Engineering Associations

UNDP	United Nations Development Programme
UNEP	United Nations Environmental Programme
Unesco	United Nations Educational Scientific and Cultural Organization
WHO	World Health Organization
WMO	World Meteorological Organization
WWF	World Wildlife Fund

1
Prelude (1905–1927)

Education starts at birth, and genes are sorted out well before that. I was born and brought up in London near the Bayswater Road, which perhaps accounts for the fact that on going to town subsequently, my first thought was generally how to get out of it soon. My father was secretary to the Institution of Mechanical Engineers in Great George Street and used to walk there and back across Hyde Park. He was expert with tools and taught me to use my hands. He came from a line of engineers and they in turn from yeomen farmers in Lancashire. Generations were long-hop, and 85 years separated my grandfather's birthday from my own. My mother was the eldest daughter of a well-heeled solicitor, J. S. Beale, who in 1895 had built Standen, near East Grinstead in Sussex, now a property of the National Trust, as a country home for his large family. As children three sisters and I spent summer holidays at Standen, where the house with its Morris wallpapers, the splendid gardens cherished by our grandmother, and the farm, of which the lower part and the young Medway river now lie under Weirwood reservoir, were an education in themselves.

In 1914 I stood at the knee of my grandfather Worthington, then 94. As a child he had been taken to the opening of the first railway from Stockton to Darlington in 1825, and had worked as an engineer right through the great years of railway expansion in Britain and also in France. He had surveyed the line of the Lancaster–Carlisle railway over Shap and, much later, in retirement, had been commissioned by landowners to *stop* the branch line to Bowness from being extended up Windermere's shore to Ambleside.

'Now me boy', he said, 'when you grow up you'll become an engineer, I expect, like your granddad, your uncle Willie, and your father, so just listen to an old engineer's advice. When you are schooling don't you worry too much about mathematics, mechanics, and that sort of thing. Go into the country and learn about trees; see the different kinds of grasses; study the rocks, the soil, the flowers, and the birds. Then when you are grown up and have learned your calculus and how to use a slide rule—you'll know what slope to make your embankments, what plants to use to hold the soil, and you'll take pleasure in your work. You'll know how to engineer with nature, not against it.' That small boy took the advice seriously: he became an ecologist instead of an engineer.

The war years, 1914–18, coincided with my spell at prepschool—at Langton Matravers on the Isle of Purbeck presided over by a notable schoolmaster, Tom Pellatt, who specialized in scholarships to Eton. However, he also encouraged butterfly collecting in that wonderful neighbourhood for bird and insect life, and that set me on the path of a naturalist. He also encouraged athletics and I walked off with the champion's trophy, following in that capacity that great athlete W. S. Bristow who in addition to a career in business became a leading authority on spiders.

A few weeks before the world war ended in 1918, two boys named Corner and Worthington entered Mitchell House at Rugby School and found they had similar interests in natural history and athletics. The association continued later at Cambridge University. One of these boys, the younger by nearly a year, was leader in many of their joint activities; he could run a bit faster, jump a bit higher, and was generally ahead in examination results. The other could swim faster, throw a cricket ball further, and he also played the flute.

Formal athletic achievement was not very important to these two, although one way and another they both represented their school and university. The first went up to Cambridge with the reputation of being the best full back that Rugby had produced for decades, but was disinclined to play with a number imprinted on his jersey, so that was the end of his football career. The other indulged in swimming and skiing. Informal athletics took many forms; climbing trees and brachiating through the branches like apes, heaving heavy stones in a competitive lob and counter-lob, slings, boomerangs, and much else. Such activities in the countryside brought these boys into close contact with many aspects of the natural world and so came the realization that both must become biologists rather than follow family traditions.

They were both avid collectors. Such well-known groups as butterflies and flowering plants tended to be neglected, for they offered little opportunity for discovery, so at Rugby, under the influence of an outstanding teacher who had worked in the tropics, mosses, beetles, and wasps came in for special attention. Later, at Cambridge, Corner turned towards botany, Worthington towards Zoology; the focus of field interest changed to fungi, centipedes, and grasshoppers. The interest was not so much in identification as in where the organism lived and what it did. For instance, having learned the songs of the different kinds of grasshopper, one could sit on a summer's day in heathland, in a hay field, or a wood and write down the list of species in that habitat and their relative abundance; or in the toadstool season the different kinds could be related to soil and climate. There was often a fry-up at the end of such a day, when the basketful of toadstools could be classified as those that were edible, those that were poisonous, and those that were eaten on the Continent. The first research paper of one boy was on the cup-fungi or discomycetes, of the other on the maternal habits of earwigs, both written while undergraduates.

At Rugby the headmaster for most of our four years was A. A. David, who encouraged individual initiatives of this and other kinds; but when David was appointed Bishop of Liverpool he was replaced by W. W. Vaughan, who was less liberal in his approach to education. This may have been good for the school at that juncture, but it was bad for us two senior boys. He sacked our biology master and all-but suppressed the science side.

At Cambridge activities were by no means limited to work, nature study, and unorthodox athletics. For the flautist there was music. Charles Wood, Professor of Music, had created the Scales Club at Gonville and Caius College, and the University orchestra under Cyril Rootham had an ambitious repertoire. Then came an absorbing interest in dance as an art form. The Cambridge Morris Men were active, ending each year with a camping tour through the Cotswolds. They came from diverse disciplines and included about that time Arthur Heffer of the bookshop, Rolf Gardiner the writer, Conway Waddington the geneticist, Alfred Peck the classisist, Joseph Needham and Robin Hill the biochemists, and Joe Coles the engineer. Traditional dance opened the door to ballet. Stimulated by Diaghelev's productions at Covent Garden, we formed a Cambridge ballet society which danced with more beef than beauty. Lokopova, married to Maynard Keynes, came to stimulate us, and so did Ninette de Valois, who was planning for her great achievement in creating British ballet. My youngest sister joined Ninette's school and had the privilege of miming Mrs Job in the first production of Vaughan Williams's 'Masque for Dancing'. But she soon married and was carried off to Northern Greenland, ballet shoes and all, by Laurence Wager, the geologist, who had also been a Cambridge Morris man.

The University Science Club in which a group of young men read papers to each other, and the Natural History Society, promoted other initiatives. I was trained as a high diver and also joined the Cambridge ski team which for the first time won its annual race against Oxford. The occasion was noteworthy for more enthusiasm than skill: Arnold Lunn started both teams in a 'schamozle' at the top of the Lauberhorn above Wengen, and we rushed down most of the way to Grindelwald. The first home, Bunny Ford of Cambridge, finished on one ski and I, running as second string, was left behind in collision with a fence post, but recovered to overtake all but one of the Oxford team. What a contrast to the modern ski race!

In 1927 both the young men mentioned above, having achieved first class in Part 2 of the Natural Sciences Tripos, one in botany, the other in zoology, were inspired by the experiences of Darwin, Wallace, Bates, and many others to work in the tropics. I had in fact been booked as a member of a projected expedition to the Great Barrier Reef of Australia, but it was deferred for a year and I could not wait. Fortunately another opportunity turned up, to assist in a fishery survey of Lake Victoria in Central Africa, whither I was off within a few weeks of final examinations. John Corner initiated his tropical career at the Botanic Gardens in Singapore. He reached high distinction in research and produced many scientific papers and books on trees, fungi, and evolutionary theory, based on his own field work in Malaysia, Ceylon, Latin America, and the Solomon Islands. He is now Emeritus Professor of Tropical Botany at Cambridge and he continues to conduct research, pays visits to the tropical forests from time to time, and cultivates his garden. His latest book (Corner 1981) records the occupation of Singapore, 1942 to 1945, from an unusual angle and couples it with detailed appreciation of two outstanding biologist-administrators, one a nobleman, the other a Professor.

In the summer of 1930, while actively engaged in arranging the Cambridge Expedition to the African Lakes, I married Stella Johnson, forming an association which lasted until her tragic death in 1978. Stella was one of five daughters of the Revd Menasseh Johnson, who did not believe in university education for girls. Therefore, after St Paul's School, where she was distinguished in athletics and music, Stella taught at Roedean for three years and introduced tennis to the young ladies, who had up to then played only cricket in summer. It was the headmistress of Roedean who eventually prevailed upon her father to let her go to Cambridge. She was able to spend only one year as an undergraduate, but this enabled her to play first string for ladies tennis, and to expand her geographical knowledge under Frank Debenham.

Marriage for a second time in 1980 to Harriett Stockton of Cape Cod, Massachusetts, added a fresh stimulus to complete this book.

2

Great African Lakes (1927–1933)

For a naturalist from temperate lands nothing could equal the thrill of entering the tropics for the first time. But think what he misses today in a journey, say, from London to Nairobi! In July 1927 there were three weeks in a British India liner: across the Mediterranean in rough weather; the smells and sights of Port Said; the heat of the Red Sea where dolphins and flying fishes could be seen from the bow of the ship; at Port Sudan the marvel of swimming over a coral reef, teeming with life of many kinds which one had learned about in lectures and laboratories; railway lines being unloaded from the ship by hordes of nearly naked Africans; night after night the Great Bear descending to the north, Scorpio dominating the sky, and after a while the Southern Cross rising. Then at last came the first sight of tropical land near Mombasa, dotted with bayobab trees which were once described by Livingstone as like carrots planted upside-down; sweating on the quay at Kilindini while checking several tons of scientific and fishery equipment for onward transmission to Lake Victoria; entraining for Nairobi.

That first tropical land journey was magical. From the mangrove-fringed harbour we traversed the narrow coastal belt of luxuriant vegetation, and as the train rose puffing towards the uplands there were entrancing glimpses of coral reefs fringing the Indian Ocean. Palm trees and bananas grouped themselves gracefully against the sunset. Then, as darkness settled, the crickets, cicadas, and frogs began their concert, and at every turn of the line exotic scents wafted through the carriage window. We dined in the Dak bungalow of a wayside station, for there were no restaurant cars in those days, and there was some delay in getting started again because the Scottish engine driver had resorted to the bar. Then in the early morning I sat in a deck chair at the front of the locomotive above the cow-catcher for a first sight of the multitudes of wild life in what was then the southern game reserve of Kenya. Yellow grass stretched away in gentle undulations to dark kopjes on the horizon; herds of wildebeest, gamboling with heads down and tails up; hartebeest lolloping along like ungainly ponies; Grant's gazelle standing delicate but stately, and little Thompson's gazelle, with their white lateral stripes and ever-flickering tails grazing unperturbed as the train passed within a few yards. A giraffe peered over a near-by thorn bush; an ostrich tested its speed against the train, and a hyaena—or was it a lion?—slinked into a reed swamp. This journey alone made those years of swotting zoology worth while.

We were bound for a fishery survey of Lake Victoria, 41 840 square km of fresh water, approaching the size of Ireland. There we had been promised the use of a steamer tug-boat, the SS *Kavirondo*, as a research vessel for six months. The survey had been commissioned by the three East African territories, Kenya Colony, Uganda Protectorate, and Tanganyika Territory, which surround the lake and could take joint action on common problems through the East African Governors' Conference.

The leader of this expedition was Michael Graham, naturalist at the Lowestoft Fishery Laboratory of the Ministry of Agriculture and Fisheries. He was then aged 29 and like myself had never before been in the tropics. He was my mentor, taught me fishery science and much human understanding and sympathy as well. Although Michael later rose to the top of his profession as Director of Fishery Research to the Ministry and trained a number of brilliant successors, the lack of any ruthless streak in his character perhaps prevented him from being elevated still higher. His

Fig. 1. Michael Graham on the SS *Kavirondo* in 1927.

books, which include *The fish gate* and *Soil and sense*, published during the war, though not widely read at the time, were precursors of the environmental revolution. After retirement Michael ended his life as a kind of Pied Piper on a pony in northern England, trailing a flock of children who helped him to convert hideous slag heaps into green tree-clad hills.

Apart from myself, the greenhorn, other members of the survey were resident in East Africa. Dick Dent, the first fish warden in Kenya, was lent to us for a good part of the time from the Kenya Game Department. He had been an elephant hunter and a coffee farmer, but above all was a born naturalist with a passion for fish. He taught me bush craft in Africa, and later in the Isle of Skye how to catch trout with a fly and to shoot grouse over dogs. Three years later we were lucky again to have him attached to the Cambridge expedition to the East African lakes.

The skipper of the SS *Kavirondo* was I. P. Stevenson, a young officer of the Kenya and Uganda Marine who had started his career before the mast under sail and had worked up to his Master's Certificate. In addition to handling his ship admirably in all conditions, he took an active part in the investigations. After the Lake Victoria survey and Michael Graham's return to England, 'Steve' was attached to me for the fishery survey of Lake Albert and Lake Kioga which we did largely without mechanical power. On Lake Albert he took one look at the 16-foot metal rowing boat which was to be our home and research vessel for several months and said 'we'll fit her with sail'. He did just that after buying up the stock of khaki drill from the one local shop, and in her we sailed the lake, up the Victoria Nile to Murchison Falls and down the Albert Nile.

Fig. 2. SS *Kavirondo* on Lake Victoria with a split-prowed canoe in the foreground, papyrus to the left and reed-swamp to the right.

The *Kavirondo*'s company included another marine officer, W. C. Barton as chief engineer, who also assisted the work in many ways, and a Goanese mate and second engineer. Most important were the complement of African sailors, firemen, and local fishermen who, on balance, probably taught us more about the fish and fisheries of Lake Victoria than we were able to teach them. One of the fishermen, Pangrassio by name, later became my head fisherman on the survey of Lake Albert and Lake Kioga and three years later was with us again throughout the Cambridge expedition. On a remote shore of Lake Edward he almost died of relapsing fever, but recovered to be employed later by the Kenya Government initiating a commercial fishery on Lake Rudolf.

After this first expedition, which for me lasted a full year until August 1928, I was so involved with the fascinating ecological problems of the African lakes that, while working on the results as a young don in Cambridge, I was busily planning a second and more ambitious expedition to the more remote and less known lakes, in particular Rudolf and Baringo in Kenya, Edward and George in Uganda. This, the Cambridge Expedition to the East African lakes, occupied another year in the field, from October 1930 to 1931, and included three other members from Cambridge.

Leonard Beadle, zoologist and chemist, had already done ecological research in the swamps of the Grand Chaco of South America. He made a special study of the chemistry of lake waters in relation to the fauna and flora and also in the adaptation of animals to swamp existence. Leonard found East Africa so much to his liking that later on he became Professor of Biology at Makerere University in Uganda where he continued with aquatic research. He trained many students who became well known in academic and other walks of life in East Africa and further afield in Nigeria and Zambia.

Vivian ('Bunny') Fuchs joined the Cambridge expedition as geologist, having just completed his final examinations in that subject, and he was able to throw considerable light on the past history of the lakes and of the rift valleys. He had been in the Arctic with one of James Wordie's expeditions from Cambridge, but took to the tropics avidly. Unfortunately Bunny was in hospital with relapsing fever for part of the time, but subsequently joined Louis Leakey's archaeological expedition which went into the field as the rest of us returned to England. He led a subsequent expedition to the southern part of Lake Rudolf but then returned to his first love of cold climates and proceeded to fame in Antarctica as Sir Vivian, Director of the British Antarctic Survey.

The other member was Stella Worthington, my wife. We married before the Cambridge expedition sailed, but she could not be with us at the start because Lake Rudolf was a closed area subject to frequent raids from the north, and no white woman had then been there. However, this problem was later overcome and she did invaluable work in surveying and making the first maps of Central Island and parts of Lake Rudolf's shoreline. She took charge of geographical data and undertook the Commissariat.

When passing through Nairobi on both expeditions the Game Department, which also looked after fisheries, was usually the first port of call. It was housed in old wooden buildings close to the Coryndon Memorial Museum, now the National Museum, which was already in those days known for its natural history section organized by the dentist and brilliant naturalist Van Someren. The Chief Game Warden was Captain Archie Ritchie, an aristocratic Scot, tall and handsome, robust of figure and with flowing white hair. He had read zoology at Oxford and during the First World War had served in the French Foreign Legion. Queenie, his wife, was the perfect hostess and they drove round the country in a yellow Rolls-Royce. Successive Governors of Kenya used Archie as a prime source of information because members of his department, including the well-trained cadre of African game guards, were *persona grata* almost everywhere, and so had eyes and ears in the African reserves and settled areas as well as in the game reserves. The Chief Warden's office door supported a necktie, the only one in the department, which served any member of it who was called to the Secretariat for consultation. On the walls were cartoons of personnel and animals drawn by artistic visitors, and in the cupboard champagne was ready whenever a pretty lady entered to take out a game licence.

Archie Ritchie's influence in Kenya, and also that of Dick Dent who occupied an adjacent office on the rare occasions when he was not on safari, was of the greatest help to our expeditions. The climax came in 1930, early in the Cambridge expedition, when we needed to take a seaworthy boat to Lake Rudolf but had no money for the purpose. At that time there was not a single boat on the lake, not even an African canoe, and our projected researches were of course quite impossible without one. There was no road to the lake shore although occasional lorries had got there after pushing and shoving across the last stretch of desert. Stella had just joined the expedition and Archie, being impressed by her charms, had an idea for solving our problem. He and Queenie arranged a dinner party at their beautiful home, where we happened to be staying, and usually dined in 'pyjamas'. But this was to be a smart affair with the Colonel commanding the King's African Rifles and the General Manager of the Kenya and Uganda Railways and Harbours as guests. Stella was placed between them with instructions to vamp them both and before the end to extract a promise from the General Manager to lend a life-boat from one of his ships on Lake Victoria, and from the Colonel to lend one of his biggest lorries and a squad of King's African Rifles in order to move the boat to Lake Rudolf, and after some months to bring it back again. Champagne flowed liberally that evening, the ladies were at their best and the men in good humour. The promises were extracted and quickly forgotten.

The following morning Archie, from his office, rang up his dinner guests. The General Manager replied to his reminder, 'Good Lord, did I really promise a life-boat? Well the Flag ship, SS *Clement Hill* is just coming into dock at Kisumu for a refit which will take several months and I suppose we might spare one of her boats for a while. I never break promise with a lady.' The Colonel, after a pungent pause replied, 'It would certainly be good practice for the troops. Will you undertake to replace the lorry if it goes over a cliff?' Thus we got our life-boat to Lake Rudolf and fitted her with limnological instruments and the expedition's outboard motor, almost the only one in Kenya at that time. The boat was all metal, well fitted with air tanks and weighed a ton. She was dubbed the *Only Hope*. Experience proved that we needed such a boat on that treacherous lake, for the seaways encountered when crossing and recrossing those waters were far too heavy for a lesser craft.

Expeditions of this kind bring home a mass of material including collections of animals and plants which need sorting and study, and extensive data and field notes recorded in log-books and photographs. Unless there is ample time to work up and publish the results much of the effort is wasted. Back at Cambridge a laboratory was provided, but there was a problem about where to live because money was in short supply. At that time we were in frequent contact with Louis Leakey who was taking archaeological expeditions back and forth to East Africa, and many were our discussions on African archaeology and zoology. Louis and his first wife Freda were living in a converted oast-house at Foxton near Cambridge, but with every penny needed for their expeditions they could ill afford the five shillings weekly rent. Since the Leakeys were about to leave for Kenya, when we returned to Cambridge we arranged to share the oast-house and play Box and Cox. Both couples could just about afford half-a-crown! The oast-house had neither water nor drainage, but then neither had a tent in Africa. It served well until babies appeared.

During these two lake expeditions a great deal of country was traversed, many lakes were studied for the first time, large collections were made for the British Museum (Natural History), and there were a good many exciting episodes some of which came fairly close to disaster. For example, near the middle of Lake Victoria on one occasion in a heavy seaway the *Kavirondo* was stopped for a scientific station and for attention to the engines. She got into the trough and all but rolled right over before the engines could be re-started and control regained. On another occasion, during a night in our sailing dinghy far out on Lake Albert , Steve and I with two Africans were caught in a storm of such magnitude that all gear other than scientific had to be jettisoned and by furious bailing we just kept afloat until Steve pointed to the tip of the spar where a glowing ball appeared: 'St. Elmo's fire: we shall survive' he shouted through the gale and deluging rain, and in due course we drove ashore and were able to sink the boat in shallow water to avoid its being pounded by the surf. On Central Island in Lake Rudolf, where Stella and I were for a week encamped in intense heat, there were some rather close shaves with crocodiles.

However, it is not my purpose to write an expedition travelogue, but to illustrate the kind of exploratory research on ecology which was being undertaken during the early years of the century by many groups like our own in various parts of the world.

The great lakes of East Africa are of two distinct types, those which occupy the rift valleys of which Lake Rudolf in the eastern rift and Lake Albert and Lake Tanganyika in the western rift are outstanding examples, and those which occupy shallow basins between the two rifts, typified by Lake Victoria. The greatest depth of Lake Victoria is 82 m, and compared with this one might expect the rift valley lakes to be much deeper since the faults which bound them plunge in many places with near vertical cliffs into what look like unfathomable depths. Lake Tanganyika indeed is extremely deep, 1470 m, the second deepest lake in the world, with its bottom well below sea level. However, the greatest soundings we made were 117 m in Lake Edward, 73 m in Lake Rudolf, and 48 m in Lake Albert.

Of the lakes investigated during our expeditions (see Fig. 3) all those in the eastern rift valley lie each in its own enclosed drainage basin with no visible outlet, for the rift valley climate is relatively dry with precipitation roughly equivalent to evaporation. With no through flow of water to keep them fresh these lakes are heavily charged with salts in solution, mainly sodium carbonate, and their levels tend to fluctuate from year to year. There is, however, one exception to this: Lake Naivasha near Nairobi which, though it fluctuates in level, is fresh water and must somewhere have a subterranean outlet. All the lakes investigated in Uganda drain to the Nile: the Victoria Nile passes through Lake Kioga, then to Lake Albert where it is joined by water from Lakes Edward and George which has come via the Semliki River, and so together, as the Albert Nile flows northwards into the Sudan. Again there is one interesting exception, namely Lake Kivu which also lies in the western rift but drains southwards to Lake Tanganyika and so to the Congo (Zaïre) River. However it was not always that way, for the valley of Lake Kivu was formerly part of the Rutshuru river system flowing northward to Lake Edward and so to

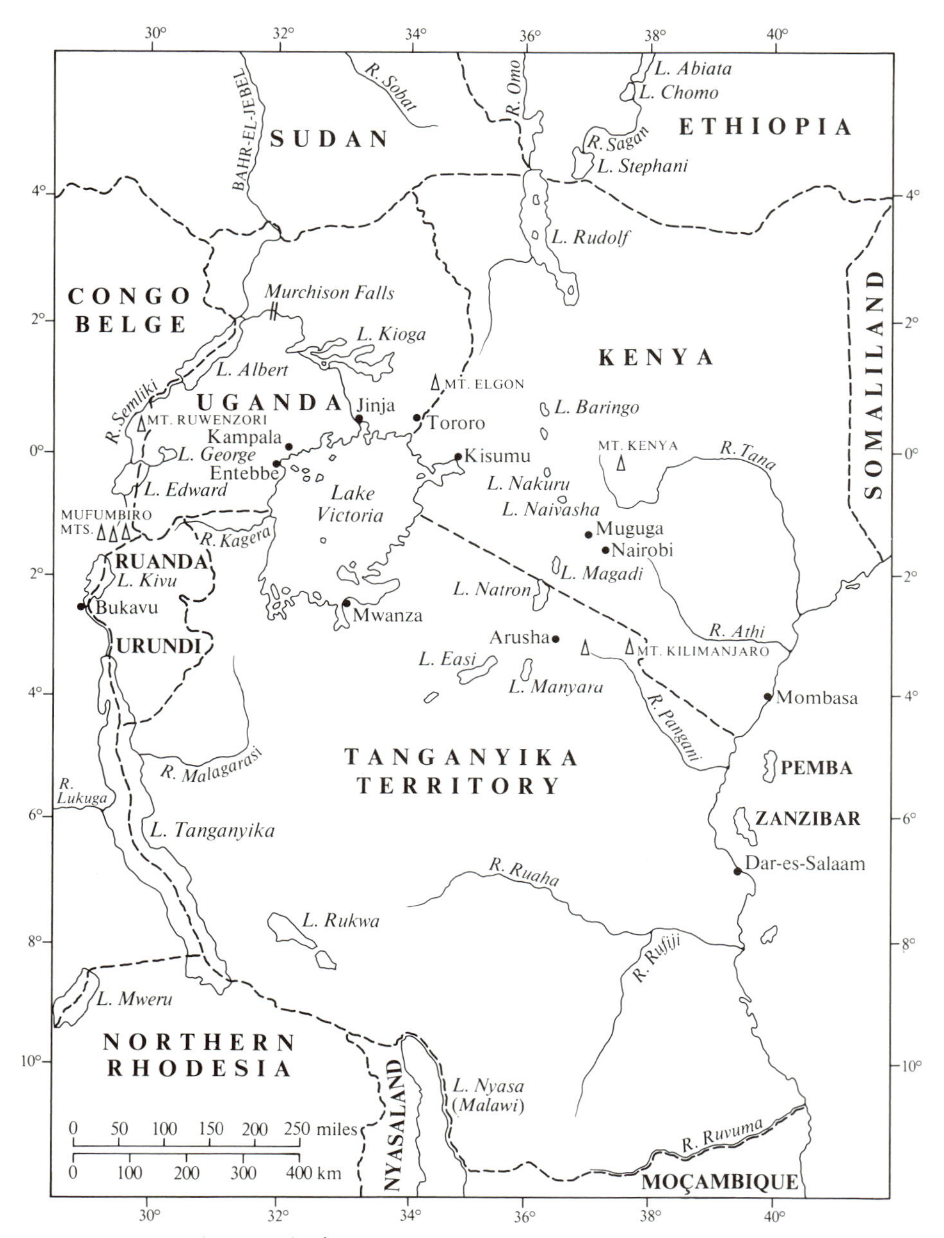

Fig. 3. Sketch map of East Africa to show places mentioned in the text.

the Nile. This valley was dammed by the range of Mufumbiro volcanos which some authorities identify with Herodotus' Mountains of the Moon. The dam of volcanoes ponded the upper reaches of the Rutchuru river until it found its overflow to the south.

In summary, what our expedition did was to carry out research on all the larger and some of the smaller lakes in this region. Advice on the existing and potential fisheries was a primary objective, but to provide this it was necessary to elucidate the ecology of each lake. Collections were made of aquatic plants and animals and these involved taxonomic studies and the description of some forms new to science. The distribution of the many different species, especially the fishes, was studied in relation to the geological history as known as that time, and hypotheses were advanced to explain this complex biogeography. For each lake the evidence from geography, physics and chemistry of the water, and the relative abundance of different elements in the flora and fauna was pieced together to provide a picture of the total ecosystem. One method of doing this was to ascertain the feeding and breeding habits of the many kinds of fish, Crustacea, and aquatic insects, not forgetting the reptiles, amphibia, birds, and mammals. A key to understanding each ecosystem and also the fisheries, was the food chain or food web, which illustrated the paths along which energy flows through the system. Mankind was, of course, added to the food chain as a major predator and his adaptation to local differences in the ecology was an important subject of study through his fishing methods. These often provided the answer how best to use the natural biological productivity to advantage.

In the rest of this chapter these subjects are recounted as they appeared to us half a century ago. Of course our two expeditions, each with little more than a year in the field, could not comprehensively cover all these subjects in all lakes. We were revealing scientific problems rather than solving them. Happily, during later decades right up to the present, many other biologists have extended these researches and some of the problems have been pursued in great detail. Much of this work has been brought together recently by Leonard Beadle (1974), by Rosemary Lowe McConnell (1975), and by Fryer and Iles (1972). The authors of those books are biologists who have personally contributed a great deal to knowledge of Africa's lakes.

Each of these great lakes contains an assemblage of aquatic animals of which some are unique. The 'endemic' forms have, it seems, generally come into being through independent evolution in that particular lake, and their number and the degree to which they differ from related species living elsewhere may indicate, in a general way, the time during which isolation from other waters has continued. Much the same applies to the flora and fauna of oceanic islands: indeed, lakes have sometimes been described as islands of water in a sea of land, especially if they have no outlet.

The interest of biologists in these lakes was stimulated first by shells which looked remarkably like marine not freshwater molluscs, brought to England from the shores of Lake Tanganyika by Captain Speke and other early explorers. This led to J. E. S. Moore's expeditions to Lake Tanganyika in 1894 and 1897 and to his hypothesis that Lake Tanganyika started its life as an isolated branch of the sea. This idea did not however fit with the geological evidence on the origin of rift valleys which was emerging from J. W. Gregory's studies about 1893, and it was finally refuted by W. A. Cunnington during an expedition led by him in 1904 which was designed to re-examine the fauna of Lake Tanganyika in the light of Moore's hypothesis. Cunnington returned via Lake Victoria from which he obtained many specimens, and a few other collections from this region had reached the Natural History Museum in South Kensington subsequently, but apart from these there was not much to go on before our expeditions. How little was known about even the commonest fishes was made clear during our early studies on the principal food fish of Lake Victoria which, ever since the opening of the railway from Mombasa to Kisumu in 1895, had been transported from Kisumu overnight to breakfast tables in Nairobi. Specimens in the Natural History Museum in London had been described as *Tilapia variabilis* but when on the SS *Kavirondo* we examined hundreds of specimens our local fishermen sorted them into two heaps which they described as '*ngege*' and '*mbiru*'. We thought the difference might be sexual, but no, there were males and females in both heaps. The accepted methods of distinguishing species of *Tilapia* gave no clue, but gradually Michael Graham and I learned from our illiterate fishermen the slight differences in form and appearance. Subsequent research showed that *ngege* had different habits from *mbiru* and lived in deeper water. It was in fact a species new to science which was named *Tilapia esculenta*. This episode stimulated a respect for the innate ecological knowledge of so-called 'primitive' fishermen, a respect which was confirmed later on when dealing with hunters, agriculturalists, and others who depend on nature for their livelihood.

The official generic name for *Tilapia esculenta* has recently been changed twice and is now *Oreochromis*, but *Tilapia* has come into such common usage that it is retained for the purposes of this book. The food of this species was found to consist almost wholly of microscopic algae living in the plankton or settling on the bottom mud in water of medium depth. With no intermediate animal link in the food chain, it is a highly efficient converter of primary production from plants to fish and it is of excellent eating quality. A commercial fishery for it with gill-nets of 5-inch mesh was started by a Dane named Aarup in 1905 and had flourished. In its early years there was a nightly catch of up to 30 fish per net, but in the Kavirondo gulf and certain other parts of Lake Victoria accessible to markets the catch had fallen off after some 20 years to about five fish per net. This sign of overfishing was in fact the basic reason for the fishery survey. Unfortunately this species is limited to a zone around the shoreline and cannot live either very close to the shore or in the open water, as we discovered from intensive netting trials in many different situations. Signs of overfishing were in fact real, so regulations to limit the fishery through the size of mesh and the supply of nets were recommended. In the events these proved so difficult to enforce that in later years right up to the present the Lake Victoria *Tilapia* fishery has become a classical example of overfishing, with a steady reduction in catch per unit effort from 30 fish per night in the virgin population to about one in the heavily fished areas today.

Recognizing that a very small fraction of Lake Victoria's potential productivity could be tapped through *Tilapia* we turned attention to a number of other fishes; there were around 200 species. How do they fit into the total ecosystem and how might they be caught commercially? About 80 per cent of the fish species belong to what was then one single genus *Haplochromis* of the family Cichlidae. These many kinds of *Haplochromis*, which have recently been grouped into several different genera under new names, rely on food organisms which range from microscopic algae through all sizes and stages of small invertebrate animals, to fishes. Their anatomy, especially the structure of the mouth and teeth, has become adapted accordingly. The majority of the species in Lake Victoria are endemic, that is, they exist nowhere else. They are closely related one to another and are sometimes very difficult to tell apart. They appear to present a *prima facie* case of a few ancestral species having undergone an *adaptive radiation* within the lake itself.

The evidence for adaptive radiation in this case is mostly from feeding habits. The many species can be grouped into those which are microphagous with small mouths and teeth, those which are piscivorous with large mouths, projecting lower jaws, and sharp teeth, and between these two an intermediate group of species which eat intermediate sized food such as aquatic insects. Another group is molluscivorous and its members are adapted to crush water snails, with pharyngeal teeth changed to a mosaic of hard flattish plates.

To the ecologist such adaptations of closely related species, known only from a single lake, appeared to be an example of *sympatric* speciation, as opposed to *allopatric* speciation which is associated with definite geographical isolation. However, leading geneticists who participated in discussion of this question, notably Ernst Mayr of the USA, would have none of this, maintaining that within a single lake it was impossible for members of one species to suffer that degree of genetic isolation which is necessary for a species to split into two or more. He maintained that there must have been times in the past when Lake Victoria, and likewise many other lakes which show a similar phenomenon, had been divided into smaller lakes which provided isolation. The mechanisms of speciation continue to be a matter of controversy for not enough is yet known about their genetic basis, but there is far better understanding now about the mode of fish life within a single lake, how the many closely related species have different ecological needs. There is a growing recognition that within one lake of varied habitat there is opportunity for ecological isolation which can prevent the flow of genes; so there is no need to postulate changes in past geography as an explanation, unless of course there is geological as well as biological evidence for such changes.

During the survey of Lake Victoria we collected not only large numbers of the many kinds of fish for subsequent expert study, but also the names by which they were known to the eight main tribes which fish around the shores and speak different languages. The multitude of different names gave further evidence of the perspicacity of simple fishermen. This study led also to description of the many fishing methods using local materials such as reeds, papyrus, bamboo, and forged iron, which were then widely practised on Lake Victoria side by side with the growing use of imported hemp nets.

The Luo tribe of Nilotic origin which inhabit the north-east corner of the lake provided particular interest in the variety of their fishing methods. Much of their equipment was made of basketwork and adapted to the particular habits of the quarry. Several of the methods involved the principle of the modern seine or purse seine in that they enclosed the fish in a large area and gradually reduced its size and thereby concentrated the fish until they could be brought to the shore or scooped out with baskets. One method represented a combination of the principle of a seine and a trawl in that the 'net' made of reeds was dragged along the bottom and incorporated a number of basketwork fish traps with re-entrant mouths so that the fish were secured in spite of the impossibility of drawing this delicate contraption on to the shore or into a canoe.

The idea for one unique method, which is used only by Luo women, appears to have been taken from pelicans. Any visitor to African bird lakes, such as Lake Nakuru, is entertained by the teamwork of pelicans driving fish by the common act of a dozen or so birds. Sometimes they encircle an area of water, then all approach the centre, and finally make a simultaneous sweep with their huge lower jaws, so that every pelican gets a mouthful of fish. The Luo women acted in closely similar fashion. Each woman with a small basket on her head carried a large fish-basket about a metre and a half across at its mouth and tapering to its centre. They waded up to their chests with scant regard for crocodiles, formed a circle, splashed towards the centre, driving fish before them, and then with one accord swept their large baskets in the water from behind to the centre and continued the movement until the baskets were inverted on top of their heads like huge hats. They continued this gambit with much pleasantry and singing until they finally came ashore each with her accumulated catch in the small head-basket.

At this time, the later 1920s and early 1930s, the subject of social anthropology was just coming to the fore under the influence of Malinowski. Much of previous anthropological science was based on material culture, and the study of the diffusion of culture among primitive peoples throughout the world had become a fascinating game, almost a cult. The shape and ornamentation of pots and tobacco pipes, the ways of tying knots, in fact almost every technology, was traced around the world. Results of such studies were of importance by indicating the past migrations of peoples, but some authorities would hardly accept that any technique could be invented locally; all must have been diffused from somewhere else. As a novice to this game, trained in zoology with its evidence of divergent and convergent evolution, I became convinced that many of these fishing methods of Lake Victoria had been invented on the spot with a minimum of copying from elsewhere. They are remarkably well adapted to the unique local conditions and habits of the fishes.

A study of the variety of canoes to be seen around Lake Victoria (Fig. 4) led to a similar conclusion, that the traditional techniques of construction had been invented rather than diffused from elsewhere. The Baganda canoe which was first described by Stanley, was already well known to students of boats. It was sometimes built to a length of 22 metres with up to 32 thwarts carrying 64 paddlers. It had a long keel fashioned from a single tree, surmounted by planks on either side sewn to the keel and to each other. The most characteristic feature however was the split prow: the keel projected several metres beyond the bow and attached to it was a curved false prow, usually surmounted by a pair of antelope horns and with a cord holding its tip to the true prow of the canoe. The origin of this remarkable feature had been much discussed. Elliott Smith had seen on the handle of a knife from protodynastic Egypt the picture of a boat with bifid stem and considered that the idea was diffused from the lower Nile southward to Lake Victoria. On the other hand James Hornell, a leading authority on primitive boats, sought the origin of the Baganda canoe in the East Indies where certain coastal boats also had bifid stems. However, after travelling in many kinds of canoe on these African lakes, where a sudden storm will often blow up rough water with short steep waves, I reached the conclusion that they formed an evolutionary series and had been invented *in situ* as adaptations to the local environment. The most

A. Blunt-ended dugout.

B. Pointed dugout, with one wash-strake.

C. Dugout base reduced, two wash-strakes and short straight false prow.

D. Upturned false prow added to *C*.

E. Baganda canoe, long upturned false prow.

F. Jaluo canoe with three irregular wash-strakes.

Fig. 4. Canoes of Lake Victoria illustrating their evolution from the dugout tree trunk to the plank-built split-prowed canoe.

highly evolved is the long Baganda canoe which demonstrated its local advantages: continuation of the keel in front of the prow gave greater stability in short steep waves, and the upturned false prow acted as a breakwater allowing all the thwarts right up to the bows to be occupied by paddlers. Other kinds of canoe used on Lake Victoria provided an evolutionary pattern which starts from the simple blunt-ended dugout made from a single tree trunk. The dugout is quite serviceable in quiet inlets and bays, but for open water it was developed to a pointed dugout with a single plank or washdrake added to the gunwale on either side and joined some feet short of the prow. The next stage was two washdrakes supported by stem and stern posts, the dugout hull then being reduced to little more than a keel, but still projecting in front of the prow. There followed the addition of an upturning false prow attached both to the projecting keel and to the prow post, and finally the Baganda canoe. When examining the most advanced stage of this series one is struck by a central groove cut on the upper side of the long beam forming the keel. This groove has no function but is reminiscent of the dugout. It seems to be a vestigial character comparable to the human appendix.

James Hornell, a sympathetic and charming character who had contributed greatly to the diffusionist theory, was quite upset at my interpretation and for some time would have none of it. However, towards the end of his life he wrote that he had come round to thinking I was right. Years later, when Uganda was looking forward to independence from the Colonial regime, the then Prime Minister reminded me about this canoe story; he had been quoting it as evidence that the people of Uganda were quite capable of inventing and doing things for themselves!

When considering the distribution of fishes in those lakes which drain to the River Nile, waterfalls are of obvious importance. From Lake Victoria there was a drop of about 6 m at the Ripon Falls (Fig. 20), which, though no serious barrier to fishes swimming downstream, was impassable in the upward direction except to large active fish such as the Ripon Falls barbel (*Barbus altianalis*). Thus the fishes of Lake Kioga, into which the river broadens some 32 km further downstream, are mostly the same species as those of Lake Victoria. Still further downstream, however, the whole river plunges over the Murchison falls with a vertical drop of 46 m to massive rocks below, and this appears to have been a complete barrier to the passage of fish in either direction. Below these falls is a highly characteristic 'Nilotic' fauna with the massive Nile perch (*Lates*), the tigerfish (*Hydrocynus*), and many other biggish fishes of diverse shapes; but this Nilotic fauna lacks the multitudinous species of Cichlid fishes which are so characteristic of the 'Victorian' fauna above the falls. Very few species of fish among the 200 or so in Lake Victoria and the 46 in Lake Albert are common to both, and the few which are, such as the lung-fish

(*Protopterus*) and some kinds of catfish (*Clarias*) can breath air and are capable of surviving long periods of desiccation dried up in mud.

The fossil record shows that typical Nilotic fauna, including the Nile perch and a large leathery water tortoise (*Trionyx*), which today are not indigenous above the Murchison Falls, once lived in the Lake Victoria basin; so the hypothesis emerged that the Nilotic fauna, which became almost extinct above the Murchison Falls was once more widespread. Many of its component species or close relatives live also, even today, in rivers and lakes of West Africa, and this indicates a former water connection across the continent in the region of the southern Sahara. Modern geological opinion is that Lakes Victoria and Kioga, which today occupy a large part of the White Nile system above the Murchison Falls, date only from the Pleistocene although there was probably an earlier lake nearby. What cataclysm exterminated most of the earlier fauna is uncertain, but from the fact that nearly every species of fish which today exists both above and below the Murchison Falls is adapted to breath air suggests that drought was an important factor.

Perhaps the original lake dried into swamps during interpluvial periods. When the new Lakes Victoria and Kioga filled, the fishes which had survived the droughts were undoubtedly supplemented by some others possibly carried in by birds. The new lakes must have been well supplied with a multitude of ecological niches, largely unoccupied, and to fill them the few ancestral fishes appear to have given rise to a speciation which produced the astonishing variety of closely related species which characterizes the Victorian fauna of today.

When we came to study Lake Edward and Lake George there was a further problem in that their fauna proved to be of the same general type as Lake Victoria. It was a part of the Victorian rather than the Nilotic fish region of Africa, though connected to Lake Albert by the Semliki River. There was a further problem that crocodiles did not exist at that time in these two lakes although they were abundant in the lower Semliki and Lake Albert. Trying to solve this problem, two members of our Cambridge expedition, Leonard Beadle and Bunny Fuchs, made a long foot safari down the Semliki River and they found a series of rapids surrounded by dense forest though not an abrupt waterfall. Fuchs had already discovered crocodile teeth and Nile perch vertebrae in the Pleistocene beds along the Kazinga Channel which connects Lake Edward and Lake George. The most likely cause of their extermination from that part of the Nile catchment seemed, as in the case of Lake Victoria, to be intensive drought during interpluvial periods, for which there was a good deal of geological evidence. The hypothesis emerged that after Lake Edward filled again the Semliki rapids provided an insurmountable barrier to the Nilotic fishes, and being flanked by dense forest, were impassable

Fig. 5. Food-chain of Lake Albert illustrated by a Banyoro method of fishing.

even by crocodiles. There must however at some stage have been communication between Lake Victoria and Lake Edward of which the faunas show some remarkable similarities.

The food chains of the Nilotic aquatic region downsteam of the Semliki rapids and the Murchison Falls are quite different from those of the Victorian region above. This was illustrated well by the Banyoro fishermen of Lake Albert who employ a particular fishing method which provided a key to the ecology of that lake (see Fig. 5). They make bundles of grass or brushwood and sink them to the bottom in 10 or 20 m marked by buoys of the cork-like Ambatch wood. A small fish (*Haplochromis albertianus*, a species which proved to be new to science) enters these bundles presumably seeking shelter from tiger-fish, and by lifting a few bundles abundant bait is obtained. Each canoe has a light fishing rod which, baited with *Haplochromis* soon catches a tiger-fish. Working up the food chain the tiger-fish, perhaps a pound in weight, is fixed on a big iron hook with a strong line tied to the canoe, and this catches a Nile perch. Thus, after adding to these three links in the food-chain, two more at the bottom and one at the top, we have an epitomy of production ecology in Lake Albert.

Piecing together all the evidence of this kind it was possible to develop a working hypothesis of the origin and evolution of the vast numbers of African freshwater fishes. The original stock of many groups certainly came out of the sea and, adapting from sea water to fresh water, penetrated into rivers and lakes. This started a very long time ago, before the separation from Africa of the New World to the west and Australasia to the east, as witnessed by that ancient group the lung-fishes (Dipnoi) of which only three persistent genera survive to this day, one each in the warm inland waters of South America, Africa, and Australia. Many of the dominant African fishes indicate their origin from the sea, for instance the Nile perch is one of the few freshwater representatives of a family of marine fishes, *Centropomidae*.

It seems that for a very long time Africa, which is one of the most ancient land masses of the world and was eroded down to a series of peneplains, must have developed rivers flowing east or west, which from time to time were more or less in contact one with another at the source, allowing the passage of fish and other aquatic organisms. At that time each river had a fauna very similar to all the rest. Even today such contact exists in the case of some rivers which rise in country which is nearly flat, for example some of the headwaters of the River Zaire flowing to the Atlantic and the Zambesi flowing to the Indian Ocean.

Then came earth movements, a down-warping of the centre causing shallow depressions which are occupied today by lakes such as Victoria; and the cracking of the surface to form rift valleys on either side of the central depression thereby cutting off the former lines of drainage. Associated with tectonic movements there was much volcanic activity which created not only the great volcanic mountains of Kenya, Elgon, and Kilimanjaro, but a multitude of craters some of which became occupied by lakes. In some places volcanoes or lava flows emerging from them blocked valleys to form dams behind which lakes were ponded. The new drainage lines, some of which flowed in the opposite direction from the old, sometimes included waterfalls.

These various events resulted in the isolation of some lakes so that evolution and speciation proceeded apace in the warm water where breeding could take place all the year round and with frequent generations high opportunity for genetic change. In some of these waters which contained large predator fish such as the Nile perch and tiger-fish, there is relatively little diversity among the smaller fishes today, whereas in those waters which at that time lacked these savage predators, such as Lake Victoria, diversity is very great. Perhaps one of the reasons for the diversity was that, with such paucity of predators, those variants which were vulnerable because they did not precisely fit their ecological niches, were able to survive and breed.

Comparatively recently, at around the time of the Quaternary ice ages, there is much evidence of gross climatic changes in tropical Africa, and these certainly had effects on the isolation and confluence of waters. For example the eastern rift valley formerly contained several very large lakes around the shores of which the Leakeys and their colleagues were finding many traces of early man; but these great rift lakes were subsequently reduced to the string of isolated soda lakes of today. The effect of this on the fish fauna is particularly revealing in the case of Lake Rudolf which our expedition found to contain a typical Nilotic fauna with all the genera characteristic of Lake Albert and the lower Nile, but in some cases the species had come to be a little different in form. Lake Rudolf has no outlet today, but a former water-level of long duration was clearly discernible at approximately 105 m above the lake surface at the time of our expedition. There is no doubt that in former times Lake Rudolf had an outlet north-westwards via the Sobat River to the Nile, and this explains its present Nilotic fauna. Recent study has shown that this branch of the Nile drew water from even further afield, from the rift valley of Ethiopia, in which Lakes Chamo and Abaya also contain typical Nilotic fishes. Their water now flows via the Sagan river to dry up in Lake Stephanie. Further south Lake Baringo has a greatly reduced fauna, but still shows some Nilotic affinities, and an extreme case is Lake Magadi, an almost saturated solution of soda and salt. One kind of fish, *Tilapia grahami* has succeeded in adapting itself to life in very hot water—up to 40 °C in the saline in-flowing streams.

With all these physical and chemical changes in the aquatic environments it is perhaps inevitable that the adaptations of fishes left some ecological niches untenanted. When men from developed countries came to live in Africa they noticed this and took steps to fill the empty niches with fishes from elsewhere. One of the earliest and most successful introductions was of trout, both brown (*Salmo trutta* from Europe) and rainbow (*Salmo gairdneri* from North America), to upland streams of Kenya. These streams have all the charm of the trout streams of Britain, but had never become colonized by indigenous African fishes except by one or two very small kinds equipped with suckers to avoid being washed away. Many reaches of the streams are isolated by high waterfalls. The indigenous fishes need warm water and trout like it cold, and it so happens that in terms of temperature the lower limit which is tolerated by the indigenous fishes (about 14 °C) coincides closely with the upper limit at which trout eggs become addled. This critical temperature is about that of the streams at an altitude of 1700 m on the equator so that above that altitude the streams are favourable to trout, below to the indigenous fish; competition between the two hardly occurs. However, the trout food in these upper cold waters was so abundant that the trout at first grew to great sizes. But later, having reacted to the favourable conditions by rapid breeding, and having eaten out the bulk of their food supply, the trout in many streams became too numerous and small. It took a good many years for the balance to settle again, sometimes assisted by fishery management.

A more debatable introduction was designed to add a large predator to the top of the food-chain in Lake Victoria. At the time of our expeditions this was being advocated by enthusiastic anglers, and it seemed a possible way to utilize the vast numbers of small indigenous fishes in that lake which were difficult to catch in bulk. We advised against such an introduction owing to the danger of drastically upsetting the ecological balance, but many years later Nile perch from below the Murchison Falls were in fact introduced to both Lake Kioga and Lake Victoria and the results are being observed with keen interest today. In Lake Kioga the total crop of fish taken annually was very substantially increased since the Nile perch were introduced. Time will tell whether this will be permanent.

Another interesting case of filling an empty niche with an alien species, though outside the region with which we are here specially concerned, is the great man-made lake of Kariba on the Zambesi. Here the river fish have in general adapted themselves well to life in a very large lake, but one of the most important food supplies for small fish, and hence for creating a productive food-chain, consists of relatively minute organisms in the plankton of the open water. Each of the natural great lakes, with its long period of evolution and adaptation, has produced plankton-eating fishes, a different kind in Lakes Victoria, Rudolf, Albert, Edward, Tanganyika, and Malawi. One of the most successful of these, a small clupeid fish called *Limnothrissa* from Tanganyika, was introduced to Lake Kariba and already, after about a decade, a new and productive fishery for this small species has developed there.

Lake Naivasha in the eastern rift valley provides quite a different example of introductions since the food-chain of today was built up from introductions. By some quirk of geological history this lake, although fresh water, contained naturally only one very small fish (*Haplochilichthys antinorii*) similar to the well-known mosquito fish. Surrounded as it was by British settlers the pressure for some useful introductions was great, so Dick Dent, as fish warden, first introduced *Tilapia* from the Tana River supplemented later by another species from the western rift valley. In this highly productive and weed-clad lake they multiplied exceedingly and have provided a useful local food supply ever since. However, *Tilapia* is not much of a sporting fish so some large-mouthed black bass (*Micropterus salmoides*) were obtained from America in the 1930s and after various vicissitudes they flourished too and now provide a tourist attraction as well as a local industry.

This was all quite successful and the great abundance and variety of birds inhabiting the reed and papyrus beds and areas of water-lilies, which are so well known on this lake, did not suffer. However, in quite recent years a very large rat-like rodent, the coypu (*Nutria*), was brought to Kenya from South America to be farmed because of its valuable fur. It escaped, and before long appeared in Lake Naivasha, a habitat very amenable because of its abundant food supply of water-lily tubers and reeds. Within two or three years the coypu population exploded and almost the entire water-lily area and much of the reed beds disappeared. The count of bird species during a few hours boat trip on the lake, which had formerly been well in excess of 100, was halved. In addition a freshwater crayfish was introduced, also from the New World and multiplied even more than the coypu. But during the last few years it seems that the population of both these recent introductions has crashed, presumably through eating out their food supply, so that ultimately, with luck, a reasonable balance may be established. This Naivasha experience emphasizes the dangers of introducing animals from one part of the world to another without very thorough investigation in advance.

With aquatic plants the dangers are even greater than with animals, as witnessed by the disastrous population explosion of the water hyacinth (*Eichhornia crassipes*) throughout most of Africa as well as other old world tropics, and of the water fern (*Salvinia molesta*) in Lake Kariba and many other places. Both of these appear to have been accidental introductions from South America, probably following the hobby of aquaria and garden ponds. They have done untold damage in blocking navigation and in obliterating, even if only temporarily, great areas of productive fisheries.

Another small lake with an interesting and in some ways regrettable history of introductions, Lake Bunyoni in south-western Uganda, was also studied by the Cambridge expedition. This beautiful lake, rather larger than Windermere, lies at a high altitude in a delightful climate. It was formed by a lava flow dam and contained no fishes at all until some were put there by enthusiastic District Commissioners. Its water is quite deep with several arms extending back into hilly country with a dense fringe of water-lilies all round. In 1931 the open water had an ecology of extreme simplicity. Multitudinous small crustacea of the plankton were eaten by a huge population of frogs (*Xenopus laevis*) which were stunted in size and were all affected by a worm parasite which rendered their eyes unusually prominent. These frogs provided the staple diet for very numerous otters (*Lutra maculicolis*) which were hunted by the local people for their skins. By that time a small catfish (*Clarias*) had been introduced and lived entirely among the water-lilies which fringe this lake. In later years the introduced *Tilapia* provided a flourishing fishery which subsequently declined, and those beautiful otters of which we had observed twenty or more at a time gambolling at the surface, were practically exterminated. I wonder what happened to the frogs!

The field work of these two expeditions was completed in 1931 and within a few years their results were published in three reports to governments on fisheries with guidelines for their development, some 30 papers published in the journals of scientific societies, articles in newspapers, and a book designed more for the lay reader than the scientist. Now, with hindsight over several decades, their influence can perhaps be assessed.

First, as one of several examples of field research in ecology about that time, they illustrated how this relatively new subject, which integrated studies of the physical environment, past history, plants, animals, and mankind, could lead to a much better understanding of tropical environments. Secondly these exploratory studies led to a pattern of development and control for the fisheries: for example, a substantial fishery on Lake Edward and Lake George was initiated and has continued with a remarkably high production ever since, and later on substantial developments took place on Lake Albert, Lake Rudolf, and other waters. Thirdly these expeditions helped to stimulate interest in similar studies and developments in other parts of the tropics. The fishery survey of Lake Victoria initiated a period of regulation which, with ups and downs, lasted until independence of the three East African countries: two of its recommendations were ultimately implemented, namely the establishment of a research centre on Lake Victoria and an interterritorial authority for regulating that lake's fisheries. That however was 20 years later (see Chapter 7).

While working out results in Cambridge I was busy planning a third expedition to the lakes further south, in particular Lake Tanganyika and Lake Nyasa (Malawi) which presented different problems from those already studied on account of their great depth. This proved impossible owing to other calls, but the ideas were soon followed up by two young lady biologists who had a penchant for African exploration. Kate Ricardo (later Bertram) who had helped me in working out the results of the Cambridge expedition of 1930–31, made the first fishery survey of Lake Malawi in 1939 together with Ethelwyn Trewavas of the British Museum (Natural History) and J. O. Borley who was that country's game warden at the time; she studied also Lake Rukwa and Lake Bangweulu. Rosemary Lowe (later McConnell) made a further intensive study of Lake Malawi in 1945–7 and went on to a distinguished career in tropical fish and fisheries. The deep and mysterious waters of Lake Tanganyika were studied by R. S. A. Beauchamp in 1938; he later became the first director of the Regional Fisheries Research Institute at Jinja in Uganda (see Chapter 7). Beauchamp found that in Lake Tanganyika and also Malawi there was a pronounced change in the water at 200–300 m. Below that the waters were permanently without oxygen and uninhabitable by any animal life which needs a regular supply of oxygen. This is not a temporary thermocline, such as is developed during summer time in temperate lakes and occasionally in some tropical lakes, but a permanent chemocline. The striking difference in water chemistry between epilimnion and hypolimnion is accompanied by a change in temperature of only one or two degrees centigrade. It is a phenomenon which was first observed by our Cambridge expedition in Lake Edward, where the chemocline was at about 60 m depth. Similar conditions have subsequently been found, not only in the very large deep lakes, but also in some deep sheltered crater lakes, but it has yet to be decided how the chemical boundary is maintained.

Belgian and French aquatic biologists also became greatly interested in the African tropics so that by the latter years of the colonial epoch many further expeditions had been undertaken and there were established centres of aquatic research, not only in east Africa but in the Belgian Congo and several countries of what was then Afrique Occidentale Française. Most of these centres continue to work under their several independent governments.

From a more personal viewpoint, study of the fishing methods in these African lakes, of canoes, and of the men and women who use them, gave me a respect for the diversity and adaptability of African peoples. Their attributes in perception, adaptation, and inventiveness gave great hopes for the future. But even in the 1930s traditional learning and technology were being lost in the pervasive culture introduced from Europe. Educationists insisted on replacing the ever-changing educational environment of nature by the uniformity of the schoolroom. I am in no sense anti-education, but many authorities, including governments, fail to recognize that evolution and development progress through diversity, not uniformity.

In later life I found that many ecological principles learned from these researches on lakes applied also to human affairs. However, we should be warned by the idea in *Pericles* 'Master, I marvel how do the fishes live in the sea?' 'Why, as men do a-land; the greater ones eat up the little ones.'

3
The African Survey (1933–1937)

Africa south of the Sahara, which is the 'Ethiopian' region of biogeography, contains a great variety of habitat and its indigenous peoples are very diverse in their physique, languages, and cultures. Nevertheless in the early part of this century many of the problems of introducing administration and development were common to the whole region. Such problems were concerned especially with law and order, epidemic diseases, agriculture production, and finance. The continent had been carved up somewhat arbitrarily between the European powers, and after the First World War development of the various Colonies, Protectorates, and Mandates was being undertaken by nationals of Britain, France, Belgium, Portugal, Spain, and of independent South Africa and semi-independent Rhodesia. The government servants, missionaries, and settlers, devoted as most of them were to their new countries and people, had adopted markedly different policies and within Africa there was not much contact between one country and the next because nearly all lines of communications had been developed independently inland from sea ports, with few internal links.

In 1929 General Smuts, in delivering the Rhodes Memorial Lecture at Oxford, spoke of these matters and pointed out that a pooling of experience could be of great benefit to all. He suggested that the time was right for a survey to be undertaken of what was going on in Africa as a whole, including the gathering of scientific knowledge.

This message fell on the ears of a group of eminent Englishmen in various walks of life, several of whom had once been known as 'Lord Milner's kindergarten', bright young men at the turn of the century who had been picked to assist in the reconstruction of South Africa after the Boer War. Soon an African Research Survey Committee was formed, the word 'research' being introduced in order to emphasize the essentially objective approach which was to be adopted and to avoid any impression that the survey was to have political overtones. When it came to publication, however, the word was dropped. The chairman of the Committee was Lord Lothian who was to become British Ambassador in Washington in the Second World War. Its members included Reginald Coupland, Professor of Colonial History at Oxford, Lionel Curtis who had been instrumental in creating Chatham House as the Royal Institute of International Affairs, Sir Richard Gregory, for a long time editor of *Nature*,

Lord Lugard, the most famed of African administrators, still of lively mind and modest façade, and Sir Arthur Salter, who had been responsible for shipping during the First World War and was then Professor of Political Theory at Oxford. There were two members with biological background, Sir John Orr, then Director of the Rowett Research Institute at Aberdeen, and Julian Huxley, then Secretary of the Zoological Society. These two were later to become respectively the first Directors-General of FAO and Unesco.

Finance for the survey was provided by the Carnegie Corporation of New York, later augmented by the Rhodes Trustees. That being assured, the Committee's first task was to find a Director, someone of authority

Fig. 6. Lord Hailey, Director of the African Survey.

and of wide experience in tropical lands, prepared to travel through Africa, and likely to be *persona grata* wherever he went. They settled on Sir Malcolm Hailey, a man of strong personality with a highly distinguished career in the Indian Civil Service and shortly to retire from his second Governorship, that of the United Provinces. They could not have made a better choice. Lord Hailey, as he was soon to become, had never been nearer to Africa south of the Sahara than Port Said; he was still full of vigor and, though very thorough in the study of any subject or situation, he was astonishingly quick in picking out the essentials. The survey of Africa did not by any means complete his career, for after its publication in 1938 he continued to study African affairs and to write many treatises up to the age of 85. It was something extraordinary to witness how a man who had no personal experience of Africa until he was over 60 years old became within a few years a leading authority on that continent, and remained so for another twenty years.

However, Sir Malcolm was at that time in such demand as a negotiator in the conferences leading to the independence of India that he could not start his enquiries on Africa until the beginning of 1935; so certain preparatory studies were commissioned in advance. Hilda Matheson was appointed secretary to the Committee. She was a lady of charm and much administrative experience, sister to the then Chief Agent of the National Trust and with invaluable contacts not only in Britain but also on the European Continent. She died all too young soon after the survey was completed and in an obituary booklet was described as like a highland pony in her native Scotland, always ready to take on additional burdens, to carry them over difficult country, and to deliver them on time.

Clearly one activity essential to the survey was a study of the progress of scientific research both within Africa itself and bearing on African problems. During the search for someone to undertake this, Julian Huxley, who had himself been on a long African journey a few years before, drew the Committee's attention to a young biologist at Cambridge who had recently returned from expeditions to the African lakes. The result was that the Committee asked me 'to prepare a report on the progress of scientific research which had a bearing on Africa south of the Sahara'. This was a tough request, particularly as I was busy working out results of the expeditions and at the same time earning a slender living by teaching zoology in Cambridge. However, Professor Stanley Gardiner, who believed that his staff and students should acquire a broad rather than a narrow scientific base, reduced my teaching duties and the East African Governors Conference provided funds for a research assistant in the person of Kate Ricardo (Bertram) to work on the expedition results, so I set about the African survey in the autumn of 1933.

It was difficult to know how to start, so I first went through the previous ten years of *Nature* and extracted anything about Africa—articles, letters, and reviews. Then, armed with some scattered and disorganized knowledge, I took up introductions to important people which were arranged by Hilda Matheson in Chatham House. In the end there was a list of almost 200 authorities who provided information or helped significantly in other ways. Most of the Imperial (Commonwealth) Agricultural Bureaux were by then established and were most helpful. Their Directors included Sir Roland Biffen in plant genetics, F. A. E. Crew and Fraser Darling in animal genetics, Sir John Russell and G. V. Jacks in soil science, R. T. Leiper in parasitology, Sir Guy Marshall and B. P. Uvarov in entomology, Sir John Orr in animal nutrition, R. G. Stapleton in plant breeding, R. S. Troup in forestry. Then there were consultations with a number who were leaders in various subjects, among others Robert Broom, B. Malinowski, Audrey Richards, and Raymond Firth in anthropology, Dudley Stamp and Frank Debenham in geography, Sir George Simpson and C. E. P. Brooks in meteorology, Hale Carpenter and Patrick Buxton in entomology, Sir Albert Kitson and Sir Thomas Holland in geology, F. L. Engledow in agriculture, Sir Edward Melanby

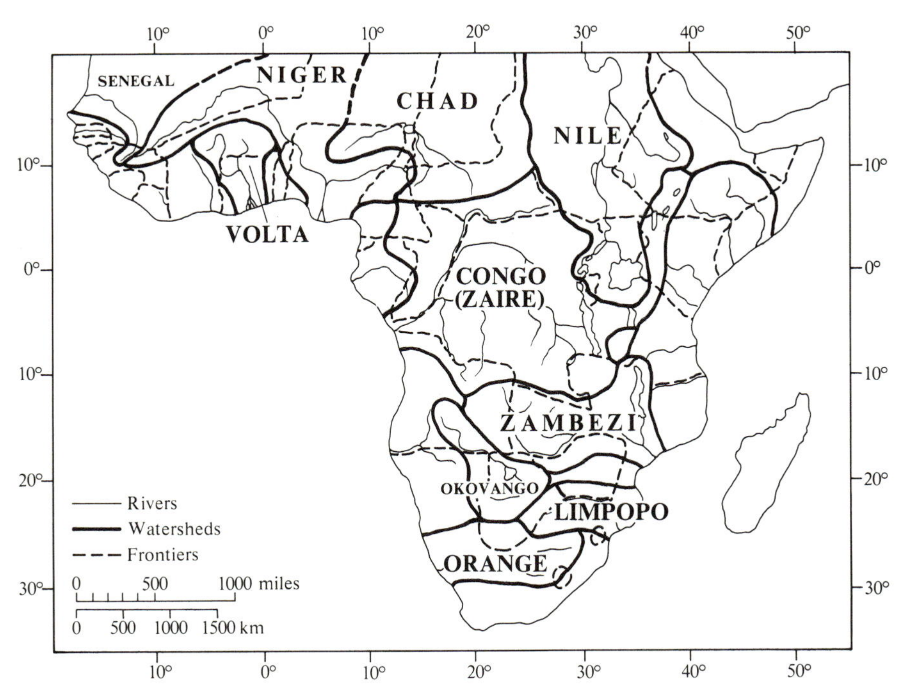

Fig. 7. River catchments compared with political frontiers. The lack of fit impedes resource planning and conservation.

and J. Rodhain in medical research, Sir Arnold Theiler and P. J. du Toit in veterinary science. Also there were specialists at the great world-wide centres for zoology and botany, Kew Gardens and the British Museum (Natural History) at South Kensington in London and their opposite numbers in Paris and Brussels.

There was a problem of how to string together all this accumulated information and opinion. The only way was to treat the whole complex ecologically; to analyse and then synthesize. There were threads running through the subjects indicating how one depended on another, so, following ideas engendered by researches described in the last chapter, I selected nutrition as a principal thread. Plant life whether domesticated or wild depended on and influenced the physical environment; animal life depended on the plants, and human life depended on both. Thus the many subjects fell into a pattern comparable to the food-chains or food-web of fishes in an African lake. Considering the rather patchy state of knowledge about Africa which existed at that time and the fact that nearly all scientific workers were involved with only one subject and often only a small corner of that one, it was important to study the relations between them. It was quite obvious that progress in one field was often seriously hampered through the neglect of related fields.

Figure 8 shows the relationships between the subjects as they emerged from considerations of this sort. Thus knowledge of the configuration of

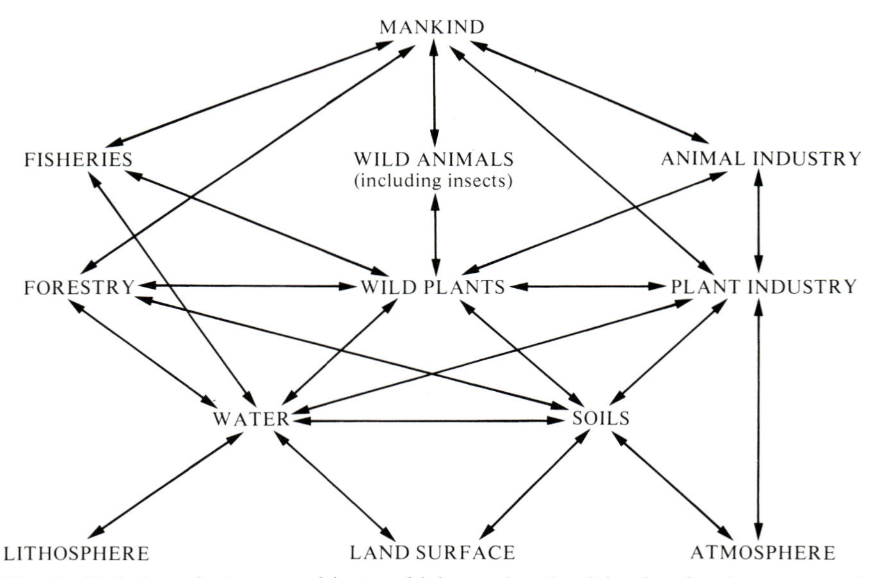

Fig. 8. Relations between subjects which are involved in the development and conservation of renewable resources.

the land (surveys and maps) necessarily precedes an understanding of the rocks comprising it (geology) and the atmosphere above it (meteorology); but the surface configuration is itself determined by geological structure and climate so that the connecting arrows point in both directions. These three subjects constitute the physical basis of the environment and between them include that all-important factor—water (hydrology). The combination of water and ground structure is responsible for the character of the soil (soil science), which in turn determines and is determined by the plants which grow upon it (botany). From the wild flora we proceed to the two main applied branches of botanical study, forestry and arable agriculture.

All the animal kingdom, whether wild or domestic, fish, insects, or man himself, depends on plants for food so that arrows in the upper part of the diagram represent food relations directly. But there are also feedbacks; the wild animals (zoology) and domestic animals (animal industry) affect the wild and domestic plants by consuming them and by manuring the soil in which they grow; aquatic animals (fisheries) have similar relations with the flora of the oceans, lakes, and rivers; insect pests and pathogens affect not only man and animals but forest trees and crops. Finally the complex of studies devoted to man himself is related to every aspect of the environment, both in his conditions of health and disease which are imposed on him by the environment, and through his activities such as agriculture, forestry, and mining which he imposes on the environment.

Thus, with a section on each of these subjects, a preliminary report on African science was drafted, and when it came to the ARS Committee Julian Huxley made a delightful remark: 'I found this report so exciting that it kept me awake at nights'. In the words of Sir Malcolm Hailey, 'The first draft appeared to the members of the Committee in charge of the Survey to contain material which merited separate publication, as a supplement to the report of the Survey, and Dr. Worthington was requested to complete his work with this object in view.' Thus my course was set. Completion of this work involved not only a most fascinating journey through a good part of Africa, but also a change in direction in my attitude to the science of ecology.

Sir Malcolm started his first African journey at the beginning of 1935 from Cape Town in two Ford motor cars, one a saloon and the other a van which carried a selection of spare parts, for roads then were by no means what they are today. To begin with he was accompanied only by Donald Malcolm, seconded as ADC from the administrative service in Tanganyika Territory, and two African drivers. The route was through South Africa, then Rhodesia, Tanganyika Territory, Kenya, Uganda, across parts of the Belgian Congo, French Equatorial Africa, and Cameroons. I was unable to join the party until that point owing to lecturing duties at Cambridge during the Lent Term, but, knowing already a fair bit about East Africa, the western side was most important. I proceeded to Lagos by sea and joined the rest of the party at Kaduna in Northern Nigeria. We spent some time in that country, then through Dahomey and Togo to the Gold Coast, including the Northern Territories. From Accra we had to take to the sea as far as Bathurst because there was no passable road through the Côte d'Ivoire and Liberia. However, our ship called at Monrovia and Freetown so we were able to see a little of Liberia and Sierra Leone. From Bathurst we motored on to Dakar in Senegal and then by alternation of rail and road to Bamako in the French Sudan and down the River Niger to Timbuktu. By then it was well into June, the height of the hot season, and the routes across the Sahara to Algeria, which were uncertain in those days even in winter time, were closed. However, Sir Malcolm, who knew well the deserts of the Indian subcontinent, was keen to see the Sahara, so we arranged special dispensation from the French Government in the Sudan and Algeria, dispatched our remaining driver home to South Africa by sea, and set off for four somewhat gruelling days northwards to reach the Mediterranean at Oran.

For the most part this journey through Africa was a somewhat sophisticated affair, staying overnight at Government Houses or with senior administrative and scientific personages at each centre. Days were long and somewhat arduous, a long drive generally alternating with a day or two based on some centre, always with a full programme of interviews, visits, discussions, and evening parties. Sir Malcolm was always keen that the work of each day be completed before the next began, and some hostesses were a little put out, after assembling a group of local people to dinner, when he excused his team as soon as eating was done in order to write up notes in their bedrooms. Sir Malcolm and Donald spent their time mostly with administrative and legal people, including African Chiefs and other senior personages, while I was with the technical staff in agriculture, medicine, and engineering. I visited many field stations, hospitals, public works, and mines, and wherever possible took a look at local lakes and rivers.

Needless to say travel by road and track in those days included a few contretemps. On one occasion Sir Malcolm, when driving the saloon himself, turned it upside down in a ditch. Donald in the passenger's seat described the next few minutes during which, in a semi-stunned condition, he felt a warm fluid running over him. How was he to get his boss to hospital? But it turned out to be engine oil and no-one, nor even the car, was badly hurt. We were often stuck in mud or sand and broke a few springs. Near the middle of the Sahara Donald and I spent half a day fitting a new half-shaft. As we worked, a thermometer under the car registered 120 °F.

The closest moment to real trouble came at the very end of the journey on the quayside at Marseilles after passage across the Mediterranean. In the Sahara both cars had given trouble in starting owing to sand in the petrol pumps, so Donald and I had developed a technique whereby one pushed the electric starter from the driver's seat and the other increased pressure in the petrol tank at the rear by blowing into it. The van refused to start at Marseilles and it was my turn to blow. I emptied a whole lungful of air into the tank without avail and while gulping another lungful petrol shot into my face. Quite a lot went straight into my lungs and, unable to breathe, I slowly passed out. I just remember a little French porter hopping about from one foot to the other crying, 'Il est déja mort; il est déja mort!', and then a pain in the chest as Sir Malcolm, the only person who kept his head, embraced me from behind with a prodigious bear-hug. By applying artificial respiration he forced out a good deal of petrol and got me more or less breathing. Later a doctor got rid of some more, but for some days the fumes were so persistent that I dared not smoke for fear of exploding.

At this time air transport was just being introduced in Africa. If the African survey had been deferred for a few years, we would probably have travelled by air from one capital city to the next and would have gained a very different impression of what Africa is really like. It was an advantage being confined to land and water. Hour after hour we drove across deserts, over uninhabited savannah, killing tsetse flies as they entered the windows, through rain forest, past plantations and areas intensely settled by African cultivators, herds of cattle, sheep and goats on open grasslands, and sometimes many wild animals. Occasionally we camped at night but we usually homed on some town or administrative centre. All this brought home the fact that by and large the most important problems of African development were (and still are to this day) in the countryside where some 90 per cent of the people then lived (and around 80 per cent of them still do so). On occasions we spent a day or two in a ship or train with time to think and write.

During the Cambridge expedition to East Africa my wife and I had a glimpse of travel to come when we moved by air from Kampala to Cairo homeward bound in the autumn of 1931. This Imperial Airways route had been opened for only a month or two and she was the first woman passenger to suffer being tossed about at low altitude for two days between Kampala and Wadi Halfa in a flying-boat, and for two more days in a Handley-Page biplane from Wadi Halfa to Cairo. A little earlier when camped at the Kazinga channel by Lake Edward we had refuelled Sir Alan Cobham's float-plane when he was exploring the possibility of extending the flying-boat route southwards down the string of rift valley lakes. That possibility was never pursued, but for several years the only scheduled air route to Uganda and Kenya was by flying boat and the sensation of descending at Port Bell, Kisumu, or Lake Naivasha was delightful, cutting the smooth water like silk. The boats were moored overnight to big black inflated buoys and there was a mystery why those on Lake Victoria were often punctured. Then a crocodile was observed in the act: the buoys were apparently mistaken for floating corpses of hippopotamus! Soon the overall decision was taken to base all major air routes on land rather than water; so every country in Africa and elsewhere had to face the problems and costs of providing ever larger aerodromes.

Results of the African Research Survey were published in 1937 by Oxford University Press. *An African Survey* is a fat volume of 1837 pages and *Science in Africa*, nearly half as thick. Lord Hailey (as he was by then) was seriously ill during part of the two years which followed the journey and were devoted to study and writing, but he was proud of the result, and rightly so. A friend of his later travelled through Africa taking with him a complimentary copy of the survey. On his return Lord Hailey asked him if it had been useful and the friend replied 'Very useful: it was just the right size to fit under the jack during repair of our frequent punctures'. But *An African Survey* was soon installed in nearly every secretariat library through Africa south of the Sahara, and on many Provincial and District Officers' shelves not only as an ornament but for intensive study. Its importance to the theme of this book is that, one way and another, it initiated a good deal of ecological thinking and the application of ecological principles. This was related not only to the biological sciences, but also to development of natural resources and to the affairs of mankind.

The survey was much concerned with the changing scene of Africa, a kind of kaleidoscope of changes at differential rates. Some of the changes were very slow, caused by physical processes; some remarkably rapid when caused by mankind, or taking place in man himself. Looking at the countryside an almost endless flat peneplain indicated millions of years of erosion by wind and water. Distant hills with flat tops rising from the plain indicated a still older base-level of erosion. With such a back-drop, there might be no obvious sign of humans, even along the roadside, because tsetse fly and the trypanosomes which it transmitted were still dominant to man and his domestic animals in the savannah. In such a scene change was minimal.

Some miles further on in a river valley there might be an area of much denser vegetation with trees close enough to provide an almost continuous canopy. Here there was no tsetse and in areas of a few hectares the trees had been felled and burned and crops were grown until the soil's fertility was so reduced that a new field had to be cut out from another piece of woodland and the deserted area left gradually to regenerate. Such 'shifting cultivation' was no bad way to make use of natural regenerative processes provided there was plenty of land to go round, but in areas of population pressure it was evident that the time for regeneration was inadequate because the soil's structure was broken down and erosion was greatly increased.

In another area verging on arid conditions climate and soil was such as to favour grasses and herbs rather than bushes and trees. Tsetse flies could not live here and there were few if any wild animals but great abundance of scruffy-looking cattle, and sheep with great fat tails as reserve against insufficiency of food and water. Here there were village settlements, but where streams and waterholes were frequent, or government engineers had established a network of water points, signs of overgrazing followed by soil erosion were evident.

In West Africa, to the north of the luxuriant rain forest belt near the coast and the savannah lands, one reaches the Sahel country which grades into the Sahara desert. Here the southern creep of arid conditions was obvious: stumps of thorn trees still fighting for survival in spite of every green branch being lopped off for foddering animals; an advancing sand dune which had partly engulfed a group of palm trees and ruined huts; goats and camels abundant but the vegetation hardly capable of supporting cattle. Indeed at the time of the Survey, nearly 50 years ago, the evidence of desertification along the southern Sahara boundary was already very clear, particularly near such large settlements as Timbuktu or Gao. An important spokesman on the subject at that time was E. P. Stebbing, then head of the Imperial Forestry Bureau at Oxford, and what he wrote in 1935 rings true today.

It took nearly two generations and several major human disasters of famine for the dangers to become widely appreciated. The big question now is whether the great amount of assistance, effort, and money being injected into the Sahel region following the drought years of the mid-1970s will do much good in the long term. Will the shelter belts of drought-resistant trees, the multiplication of water points for domestic animals, of irrigation schemes for crops, of health services to improve human survival, arrest the advancing Sahara? Or will they make matters even worse for a larger population of humans and domestic animals during the next run of drought years?

Let us look at some of the scientific subjects which are of importance to development, as they appeared nearly half a century ago. Development of any kind is not at all easy without accurate and adequate maps—small-scale for the overall picture, medium-scale for topography, and large-scale for property boundaries. In the early 1930s the mapping of Africa was nothing like it is today, because it depended on the individual surveyor on the ground with theodolite and plane table, not on the aeroplane and the computer. But either method depends ultimately on the sheets of information being linked to a network of fixed triangulation points. At that time the first-order triangulation along the 30th arc of meridian, which runs from the Cape to Cairo, was incomplete, but it was badly needed as a kind of backbone for all the secondary triangulation, much of which had been undertaken at the beginning of the century in order to establish international frontiers. Even these frontiers were floating back and forth, so to speak, until all the surveys could be tied together by the 30th meridian. The job had in fact been started from the Cape through South Africa during the early years of this century, continued through Northern Rhodesia by 1907, and then in 1931–3 northward to the boundary between Tanganyika Territory and Ruanda Urundi. From there was a gap with no geodetic control except for the Uganda–Belgian Congo frontier, as far as Wadi Halfa where Egypt joins the Sudan. Thus there were two chains of first-order triangulation, one stretching northwards from the Cape, the other southwards from Cairo, and a bit in the middle, but their ends were wobbling. In the event, the last gap in the Sudan was not completed until 1954 by an American expedition provided under their technical assistance programme. So at last it became possible to adjust the accumulated errors which are inevitable in triangulation surveys even if, as in this case, it had been of very high accuracy. With the aid of additional triangulation along the arcs of parallel at 12° north and 6° south it became possible to relate accurately one to another the surveys and maps of most of the countries in Africa.

These basic international operations allowed the topographic surveys by air photography, which were conducted years later, to proceed with greater confidence, and also the large-scale work right down to the cadastral surveys which are essential for many forms of development, from the alignment of a road to the siting of a school or a clinic.

Precision in boundaries on the ground, whether between countries or small plots, can be particularly important in the mineral industry, and this introduces the subject of geology. For every country in which a geological survey had been started at that time the development of mineral resources was its primary objective and justification. There was not then a great deal of mining outside South Africa, the Gold Coast, and parts of the Belgian Congo and Northern Rhodesia. However, there were a good many indications of resources elsewhere and small-scale mining of gold and other minerals of high value was undertaken. There were occasional gold rushes such as that to Kakamega in Kenya in the early 1930s but most of the laborious work of geological mapping paid off much later with developments such as oil in Nigeria and copper in Uganda. Of underground resources water is, however, the most important in Africa, and hydrogeology was already revealing aquifers, many within reach of shallow boreholes which did not need complex machinery to bring water to the surface.

The availability of water controls all plant and animal life, so a particular study was undertaken of rainfall records. The remarkable symmetry of the African climate was at once apparent. The tropical belt with its double climax of precipitation near the equator, following passage of the vertical sun, grades to a single climax around the tropics of Cancer and Capricorn. Then proceeding south as well as north there is increasing aridity, with its climaxes in the Sahara and Kalahari deserts. These in turn are flanked by the marine influences which cause Mediterranean type of climate and vegetation at both northern and southern extremities of the continent. Of course there are many local variations of this symmetry, caused especially by the great width of the continent north of the equator and its tapering to the south, and by local features such as rift valleys and high mountains. But the overall picture dictates to a large extent the soils, vegetation, and potential for land and water use. Not only the surrounding seas but some large areas of inland water have significant effect on climate: for example, the reliable rainfall of Uganda is due in part to the influence of evaporation from Lake Victoria, and the tea-growing uplands at the northern end of Lake Malawi, are much influenced by that lake.

During the fishing survey of Lake Victoria a few years earlier the effect on climate of that great sheet of water was abundantly clear. Except when influenced by major atmospheric disturbance, such as movement of the intertropical convergence zone, there was a daily breeze from the lake and a nightly land breeze as the land heated under the sun and cooled at night. It caused an apparent 'tide' of up to a foot at Kisumu for instance, and where high land was near by the moisture-laden lake breeze could be seen to give rise to daily cloud.

In the 1930s there was much discussion on the effects of long-term climatic change in Africa and particularly on the likelihood of progressive desiccation, but an analysis of evidence from climatological records, meagre as they were, gave nothing to support it. We concluded provisionally that the undoubted high agricultural productivity in the 'granary of ancient Rome' of north Africa, and in countries bordering the southern fringes of the Sahara, was man-made. Of course if one goes back many thousands of years there is plenty of evidence for much higher precipitation. There was undoubtedly once a water connection from east to west across the southern Sahara, as proved by the present distribution of fishes, and we picked up pieces of fossilized timber near the middle of the Sahara where no tree could be found for a hundred miles or more in any direction. The argument on progressive desiccation still continues to this day, but I do not think that a consensus among scientists would differ much from that concluded 45 years ago, namely that the climate of today, unstable as it may be from year to year and decade to decade, has been much the same for several millennia.

Another question then much discussed and still with its advocates was dowsing, for minerals as well as water. It had a particular exponent in Sir Albert Kitson, then Director of Geological Surveys in the Gold Coast, who used a kind of 'magic box', which in his view was based on scientific principles, instead of the traditional wand of hazel. For many people of scientific inclination I think the argument ended at least temporarily during the Second World War when professional dowsers were still employed, though not officially recognized, by the allied armies in north Africa: the percentage of successful finds of water by qualified geologists was much higher than by dowsers; the percentage of success by ordinary army officers who had developed a good eye for country approximated to the latter. The recent recrudescence of interest in extrasensory perception may well have given renewed zest to those who experiment with dowsing, so perhaps we shall learn more about it soon.

Proceeding in the ecological sequence from the physical factors of the environment we reach the soil, or better the soil and vegetation taken together, because, while the soil in a general way controls the vegetation which will grow upon it, the vegetation has itself a big influence on the quantity and quality of the soil. This was well known to the 'primitive' shifting cultivators who select their plots of woodland for 'slash and burn' according to the kind of vegetation which had regenerated over the years. A tragedy in modern times is that many would-be agriculturalists in the tropics do not seem to realize the dangers of wholesale destruction of woodland in the tropics. Expecting to cash-in on rich soil, they find that, once exposed to the elements, it soon loses its fertility and structure and becomes subject to serious erosion.

By the early 1930s it was already widely recognized that natural forest and other wild vegetation, especially in upland areas near watersheds, was of great importance in maintaining the flow of springs and streams in the country down below. To chop down forests wholesale and to cultivate steep slopes was the quickest way to obliterate water supply in dry seasons and to cause destructive floods at times of heavy rainfall, thereby changing perennial streams to something like the Wadis of North Africa.

There was a good deal of argument at that time about the effect of forests on local rainfall and indeed this question is not finally resolved even today, though it is known that the effect if any is small. There was also argument developing on the benefits or otherwise of the growing practice of replacing natural forest over large areas by pure stands of exotic conifers, particularly species of pine from North America. Certainly this could produce much more usable timber or pulp wood in favourable conditions but it had not the same regenerative effect on the soil, and some claimed it caused greater evapotranspiration and hence reduced percolation. Questions of this sort take many years to resolve by experimentation in the laboratory and the field. But foresters, from the nature of their calling, think a long time ahead and fortunately in nearly every country they were able to secure significant areas of natural vegetation as forest reserves before the demand for more agricultural land became too great. Most countries aimed at at least 10 per cent or 12 per cent of the total land being retained under forest though in by no means all countries was this achieved.

Forest reserves became the first conservation areas for wild animals as well as plants, but big game was such a special feature of Africa and by the 1930s was being destroyed so indiscriminately that game reserves were already established as a part of national policy, especially in East and Southern Africa. Predators such as lion, leopard, and wild-dog were first classed as vermin and shot on sight whenever there was opportunity. But soon ideas of conservation of the complete range of wild life got established and game departments became an important part of government for many years. Conservation societies in Europe were already an impetus in this movement; an example was the Society for the Preservation of the Fauna of the Empire which had been formed in Britain by a group of prominent big game hunters known colloquially as 'the penitent butchers'. In America the Audubon Society was beginning to have influence, and President Theodore Roosevelt in 1910 had pronounced the dictum of 'conservation through wise use'.

Another aspect of wild life of great importance in the total ecological picture was the food it provided for mankind. Traditional native hunting practices were in general exempt from game laws, but it was some time before it came to be appreciated that a high proportion of animal protein in the diets of many African peoples is provided by wild animals. This is particularly so in West Africa where big game is far less conspicuous than in the east but 'bush-meat' in the form of smaller mammals like cane-rat has always been, and continues to be, an important food constituent. Many wild plants too provide important food: for example, that useful tree the baobab, in addition to providing bark cloth and water stored in its carved-out boles, has leaves which are crushed and eaten in soups. Analysis showed that these leaves have high calcium content and also vitamins which are destroyed if dried in the sun. Some West African tribes take the precaution of avoiding direct sunlight when preparing their soup, which says something for their unwritten knowledge of nutritional chemistry.

But in spite of its many values, the Ethiopian flora and fauna, which contain a very high proportion of species which are found in no other part of the world, had relatively few friends. In particular the powerful veterinary departments which were successfully overcoming the ravages of diseases in domestic animals were at that time antagonistic to large wild herbivores. These African mammals had inborn resistance to the African disease organisms, but the domestic animals had no inborn resistance and were apt to succumb in large numbers to diseases such as East Coast fever and *nagana* (trypanosomiasis). There was also the exotic disease rindepest which, accidentally introduced from the East, had swept southward through Africa towards the end of the nineteenth century and had decimated cattle herds and many of the wild herbivores. Its spread through most of the continent had certainly been assisted by the migration of wild animals, particularly buffalo, which are very susceptible to this disease.

The leading veterinary research centre of Onderstepoort in South Africa had been directed by that great Swiss veterinarian Sir Arnold Theiler who had revealed the causes and methods of prevention of many of the worst diseases. It was a great privilege for a young biologist like myself to meet him for an hour or two and to learn at first hand of his researches. Rindepest, however, was primarily the concern of the laboratory at Kabete in Kenya because South Africa, where this disease had by then died out, dared not risk having the virus within its territory. In East and West Africa outbreaks of rindepest continued and the accusing finger was pointed at the game reserves as reservoirs of the virus for transmission to cattle. However, at about that time persistent and brilliant research by the veterinarians at Kabete produced a successful vaccine and soon there was an international system for its distribution throughout tropical Africa. Within a few years it was possible, with supplementary centres of vaccine manufacture as at Vom in Nigeria, to immunize all cattle in areas which were at risk.

Of the changes which were taking place so rapidly at the time of the African Survey, those in agriculture were among the most important. Agricultural officers who were endeavouring to raise the level of peasant farming and production, and who were well trained at agricultural schools and universities in Europe, had to change many of their ideas when they came to Africa. The ecology of land use could only be learned on the spot and the environment had a way of kicking back at mistakes.

Consider, for example, the practice of shifting cultivation, which has already been mentioned as prevalent in various forms all over the continent. Agricultural training in Europe had been directed towards settled agriculture, the maintenance of fertility by alternation of crops, and the use of manures whether natural or artificial. By contrast shifting cultivation, with the labour it involved for the temporary cultivation of a few acres, appeared to have minimal productivity. However, when 'improved' methods were introduced time soon showed how rapidly a fertile tropical soil can be worn out by successive croppings. Another example was the practice of mixed cropping which was traditional. Many were the occasions when a newly arrived agriculturalist would persuade the local peasants to plant a pure stand of one crop or another, only to find that some pest or disease swept through the whole whereas on a near-by traditional mixed plot the pest, though attacking individual plants, had difficulty getting through the adjoining vegetation to its next host.

Much greater troubles came however from large-scale monoculture, particularly of rice, for the bulk of the crop was apt to be consumed by swarms of locusts or millions of small finches *Quelia quelia*. In the African savannah *Quelia* is a common bird which goes about in parties of a few dozen from one patch of grass to another feeding on seeds. It has always been a nuisance in cultivation, but on a small farm its depredations could be reduced to a minimum by scaring. This provided occupation for the children with intriguing devices of strings and clappers, so that one small person sitting on a raised platform over the crops could keep an acre or so free of trouble. However, when it comes to thousands of acres and the children are at school instead of in the fields, this is impossible. Large-scale farmers have to resort to gathering the grain crop before maturity, or to poisoning the birds or to blowing up their roosts where millions accumulate at nightfall. *Quelia* remains an important factor in planning new settlement schemes and 'agri-business'.

When travelling through Africa every ecologist is impressed with the importance of fire as a dominant factor which shapes the vegetation and hence the whole terrestrial environment. It would hardly be an exaggeration to say that nearly every part of the continent which is burnable is burnt over, if not every year, at least one year in three of four. The only parts which are not burnable are deserts and forests with closed canopy which prevents any significant amount of ground vegetation. Hunter–gatherers burn the country in order to move about with ease, especially when seeking honey and driving game; pastoralists burn to discourage bush, to encourage new grass when the rains come, and to destroy ticks and other vectors of disease; cultivators burn to clear vegetation and for the fertilizing properties of ash. In the plantation industries precautions against fire are of course taken, though not always successfully; but broadly speaking a very large part of the African continent is burnt over so frequently that the wild vegetation has become a fire climax, composed only of species of which the parts above ground can survive fire, or which have roots capable of regenerating new shoots when the fire has passed. The importance of fire can be appreciated today even by those who travel in aeroplanes: from 30 000 feet in the absence of cloud one can see the ground clearly during the wet season, but in the dry season smoke in the atmosphere renders it almost invisible, except at night when the fires are all too obvious creeping across the grassland or savannah. The

Aristotelian elements of earth, air, fire, and water represent a good summary of the main ecological factors in Africa.

Although there are well-proved cases of fires being started by natural causes such as lightning, the burning is practically all man-made and is not of course limited to Africa. Taking the tropics and sub-tropics as a whole the annual wastage of organic material is very great; it would be interesting to calculate how it compares with the consumption of fossil fuels in temperate lands. But under the present extensive systems of land use (some would say land misuse) in Africa fire remains an essential tool when it is applied under control. Thus in the cattle ranching areas of Rhodesia the country is fenced into large paddocks which are burnt on a four-year rotation. This is an effective method of depressing bush growth, which, without fire, would replace the grassland. This example, and there are many like it, raises the question whether it is desirable to burn early or late in the dry season. Burning late when the herb layer of vegetation is high and dry is extremely destructive of all woody vegetation, whereas with early burning the fire is much less fierce. Thus in a general way late burning favours the animal industry by maintaining grassland, early burning favours forestry. Such questions were frequent talking points during the African survey; when foresters and veterinarians met over sundowners, the argument could become nearly as fierce as the fire.

Domestic animals have been a mainstay of human life and activity in Africa since cattle, sheep, and goats were brought from the north-east several thousand years ago by the Hamitic immigrants. It is a surprising fact that, considering that man originated in Africa, as now widely accepted, and by emigration to the Mediterranean, Asia, Europe, and later to the New World, gave rise to all the great civilizations, man in Africa never domesticated local animals. In spite of the richness and variety of the Ethiopian fauna he had to wait millennia until his descendants came back to Africa and introduced the Asian animals. But once in Africa cattle, sheep, and goats appear to have spread rather quickly in spite of diseases and the abundance of large predators, except to areas of dense forest, pure desert, and savannah infested by tsetse fly. Many are the local races adapted to different ecological conditions which have been studied and described by specialists in animal husbandry. However, with few exceptions the local races have a productivity of meat or milk, or capacity for draught, which is low by standards in other parts of the world; so at the time of the survey much effort was being put to introducing bulls of exotic breeds, mainly Zebu from India but also some of the milk and beef breeds from Europe, for crossing with the native scrub cattle. Often this was successful, especially in the early stage of hybrid vigour, but it raised still more problems of disease control and particularly of nutrition.

Animal industry in a large part of the continent tends to be of the pastoral nomadic kind adapted to the vagaries of climate and availability of grazing and water, but nomads are not easy to control, to tax, to educate, or to cure of disease. Thus, from the earliest stages of development efforts were made to persuade nomads to settle down and even to cultivate land. By no means all these efforts were successful, nor are they today. Nomadic pastoralists are quite good ecologists with an eye for country, a clear idea of what is good for them, and a resistance to change. Quite recently in Masailand I met a couple of elders in traditional dress of red ochre, blanket and spear, outside a manyatta with its circular thorn fence and low mud huts. Not speaking Masai I addressed them in Kiswahili hoping that they might understand. They answered in perfect English and it transpired that they had been through primary and secondary school in the local town, had looked at life among settled cultivating people and had decided that it was not for them. They returned to their cattle and nomadic life, and perhaps in this they showed wisdom.

However, some nomads were prepared to settle and of course many tribes are primarily or wholly cultivators by tradition, so the problems of arable agriculture were extremely important to the Survey. Pressure of population and consequent deterioration of soil were already apparent in some areas. An extreme case was the Owerri Province of Southern Nigeria where the agricultural population had already reached between 300 and 400 per square mile by 1930, and the resting period which the soil received under shifting cultivation had been reduced almost to nil. When travelling through Senegal, much of which was given over to the production of groundnuts, the reduction of soil fertility through frequent cropping was already clear. In that country three belts could be defined in terms of rainfall, one in the dry lands to the north, the second stretching east in the latitude of Dakar, and the third in the damper country surrounding the Gambia river. To begin with, population concentrated in the northern belt which was found most suitable for ground-nuts, but by the mid-1930s nearly all the exports came from the middle and southern belts. Later the focus was even further south as the areas of relatively low rainfall lost their fertility through persistent cropping.

Against such factors agricultural scientists were trying to establish systems of fixed cultivation. Rotation of crops, mixed cropping, green manuring, composting, and especially mixed farming were being designed to put fertility back into the soil as fast as it was drawn out by crops. Here again the adaptability of people who live near to nature was sometimes remarkable: an interesting case came to notice on an isolated island called Ukara in Lake Victoria which had hardly been visited by an agricultural officer. Under strong pressure from an increasing population a genuine system of mixed farming, of the type which agricultural experts were attempting to popularize in both east and west Africa, had apparently been evolved spontaneously.

Certain insects of high economic importance were at that time attracting a good deal of scientific research. Some, which were vectors of disease, fed on blood or the sap of plants in the adult stage; others, which caused damage by the quantity of vegetation they consumed, were herbivores throughout life. An example of the former, namely tsetse fly (*Glossina*), has already been mentioned but it is well to note that there are no less than 20 different species of this genus in Africa, each requiring different ecological conditions. In the 1930s the most important tsetse work was being undertaken in Tanganyika Territory under C.F.M. Swynnerton who had gone there originally as Chief Game Warden but subsequently built up a brilliant team of entomologists and ecologists who laid the foundations for tsetse control throughout Africa. Swynnerton himself was tragically killed in an aeroplane accident but two of his sons continued the family tradition, one becoming a Chief Game Warden in his turn and the other having a distinguished career in tropical agriculture. In spite of all the research work and field operations devoted to clearing Africa of tsetse fly and thereby allowing its settlement by people and domestic animals, there are still very wide areas where the fly remains dominant. Many of those who wish to see the wonderful wild life of Africa surviving for ever have a sneaking suspicion that tsetse flies may be their best friends, for they have locked up areas of wild life against the invasions of man. There is some hope that, by the time science makes it possible to eliminate the flies everywhere, man in Africa will have reached a point in development at which he gives a full and rightful place to the conservation of nature.

A key problem was how *Homo sapiens* could himself take benefit from this vast ecological complex which was Africa, how he could live and multiply on the income of the natural resources without destroying their capital (except in the case of minerals), and how he could conserve the values of Africa for future generations, not only the economic values but also the scientific and ethical values. At this time anthropology was just coming under the powerful influence of Bronislaw Malinowski. He had published his treatise on the Trobriand islanders under the title *The sexual life of savages* which became a best-seller. Of the two driving forces which seem to make mankind tick—survival of the individual and survival of the race—or, to put it more crudely, hunger and love—he had indeed focused attention on the latter. This was followed by others of the new generation of social anthropologists, so much so that Lord Hailey, after talking to some anthropologists, remarked to me in a cynical mood 'No anthropologist is considered worth his salt until he had discovered a new obscenity.' The more mundane aspects of human behaviour, concerned with getting enough to eat, were studied with rather less enthusiasm until

Audrey Richards published her study of the Bemba in Northern Rhodesia under the title of *Hunger and work in a savage tribe*. The thoughts expressed in that title provided the most obvious links between mankind and each ecosystem of which he happens to be a part, and also help to link together the many aspects of his environment.

In British Africa the policy of indirect rule, which had been initiated by Lord Lugard in Nigeria, had spread very widely and this stimulated many studies analysing those institutions, economic, political, and legal, on which the everyday life of Africans is based, and the changes which these were undergoing in response to the many agencies of European contact. In this Malinowski was certainly a dominant figure although his field-work had never been in Africa itself. To obtain some understanding of the subject I used to attend some of his seminars at the London School of Economics, where incidentally one rubbed shoulders at that time with Jomo Kenyatta and other students from Africa who later became prominent in social and political fields.

Most of the studies on the human species, however, had been not so much on its relationship to the external environment, as to its internal environment, in health and disease. Medical science had indeed outstripped behavioural and social science, and its technologies were already being applied in quite a large way in Africa. Initially the focus of both government and mission doctors had been on curative medicine and establishing hospitals, but discovery of the causative organisms and vectors of such widespread diseases as malaria and sleeping sickness at the beginning of the century was leading to the conclusion that a greater service could be rendered by prevention than by cure. However, in the 1930s, before the discovery of synthetic drugs and insecticides which came to dominate the scene after the Second World War, there was a good deal of argument how best prevention could be achieved. Take malaria for example: in many parts of the region nearly every African was infected before he was three years old, and if he did not die, developed at least a partial immunity. Measures were being taken to stamp out malaria, particularly in townships, by drainage through engineering works, treating static waters with the insecticide Paris Green, and by sanitary inspections. A section of medical opinion was favouring the extension of such measures into rural areas, but another section pointed out that, while this would reduce the very high infant mortality, the adult African, moving from one place to another, would be almost certain to be infected later when his physiological adaptability was reduced. Later of course, with the discovery of the synthetic insecticides and drugs during the Second World War, very wide-scale use of sprays and of new anti-malarial drugs practically wiped out the disease in some parts of the continent, albeit at the expense of damage to the environment. Now it looks as though the wheel is coming full circle as a result of genetic resistance to the synthetic chemicals, developed both by the pathogens and by their insect vectors, so that serious outbreaks of malaria are taking place in many areas from which the disease had supposedly been eradicated. To keep one jump ahead new synthetic formulations are frequently needed, each of which may have a relatively short commercial life, and each new formulation costs millions of dollars before it can be put to use. Thus recent discussions on controlling malaria and other water-related diseases have focused attention once again on the engineer and sanitary inspector as the long-term solution. After all, malaria, once rife in the English fens, was not eradicated by toxic chemicals but by drainage and agriculture (see Chapter 11).

In the 1930s yellow fever, which is transmitted by the mosquito *Aedes*, was also often contracted by Africans at an early age, when the patient either died or attained immunity for life. Intensive research, mainly by the Rockefeller Yellow Fever Institutes in Uganda and Nigeria, which had caused the untimely death of several brilliant research workers, showed that this disease was not restricted to the endemic centres on the west coast, but extended in dormant form throughout large tracts of tropical Africa. With the development of air communications the period of incubation of the disease was longer than the time of travel, so there was danger of its spread to huge populations in other continents where yellow fever was unknown. Once inoculation became possible on a large scale, and with diligence in the insecticidal spraying of aircraft to kill any mosquitoes which might have gained a passage, such a human tragedy was avoided; and now yellow fever is almost a disease of the past. It is certainly a great success story of preventive medicine, together with smallpox which was declared by WHO to be an extinct disease in 1980.

It is strange to think now in the 1980s that the problem of human population explosions was hardly considered by the African survey. There were of course situations of overcrowding, both rural and urban, but they were comparatively rare and localized 50 years ago. They did, however, lead to thoughts of what the situation might become a couple of generations later when health services would have their full effect in increasing the survival rate of babies and lengthening the expectation of life. Where reliable vital statistics had been collected, infant mortality rate was in some areas as high as 40 per cent and the expectation of life at birth less than 30 years. How long would it take for some populous areas to burst at the seams with too many people when the infant mortality was reduced say by a half and the expectation of life perhaps doubled? When I mentioned such qualms to Lord Hailey during our travels he would have little to do with them. He was impressed with the underpopulation of Africa as a whole and, from long experience in India, pointed out how progress and development had come only when the population increased.

In his view a bit of human pressure on land resources was no bad thing in encouraging development. However, he was ever prone to argument based on proved fact and when, not many years later, A. V. Hill returned from a visit to India well primed with statistics on the extreme rapidity of population growth and the real dangers of a population explosion before education, religion, and the general economic, and social level could result in family planning on a wide scale, Lord Hailey was quickly converted. He soon realized that, without some form of population control, there were a good many parts of the world which would soon be heading for disaster.

This whole survey of science in Africa attempted to depict a cross-section of activity as it was in the middle 1930s, but the cross-section was possible only by imagining the process of change to be suspended for examination. The real picture of Africa at that time as indeed at the present, was not a still but a cinematograph, with all branches of physical, biological, and human activity changing, singly and collectively, and reacting on each other. This was the essence of ecology.

In this chapter we have looked at only one part of the African survey. Much of my scientific assessment was incorporated into Lord Hailey's big volume but the survey covered much else: systems of government, law and justice, administration, taxation, labour, mines, transport, and communications. It had a considerable impact, especially on governments which were at that time responsible for different parts of Africa south of the Sahara. In Britain it led to the Colonial Development and Welfare Acts which reached the statute book during the Second World War, and these included the allocation of a proportion of the funds specifically for scientific research. This in turn led to the creation in London of a Colonial Research Committee, and later indirectly to the Scientific Council for Africa South of the Sahara (CSA). Thus it fell to my lot to conduct a second survey of science in the development of Africa just a quarter century later than the one here described, but that is another story (Chapter 8).

4
Ecology in the Lake District (1937–1946)

Early in 1937 there appeared an advertisement for a director of the Freshwater Biological Association of the British Empire (FBA). This resounding title was attractive, particularly in suggesting contacts overseas, but I was reluctant to apply because close association with Lord Hailey had widened my interests to a point at which my special subject of limnology was a bit limiting. In the end it was Julian Huxley who pursuaded me to apply. I had been in touch with him at intervals and had learned to heed his advice ever since he examined me for a scholarship to Oxford which did not come my way. During undergraduate times he became interested in my studies of earwigs in connection with disharmonic growth. Later he was intrigued with the speciation of African fishes, and latterly we had many dealings over the African survey. His voice was persuasive.

So later in 1937 I had an office and laboratory in a great sham medieval castle situated on Windermere (Fig. 9), with my family ensconced in a beautiful house and garden outside the village of Hawkshead. There

Fig. 9. Wray Castle, headquarters of the Freshwater Biological Association until 1947, with the north basin of Windermere beyond.

followed nine years based in the Lake District devoted to developing the scientific potential of the FBA (the Empire bit in its title was dropped after a while). From 1939 to 1945 the work had, of course, to be adapted to the war effort.

In order to appreciate how this episode fits into the ecological century we must glance at the development of freshwater biology, or limnology as it is often rather misleadingly called. Properly the term limnology should be limited to the study of lakes, and can reasonably be stretched to reservoirs, but it has come to include all inland water bodies, including rivers and streams. Even freshwater biology is not an adequate description of this branch of science for the quality of inland water bodies ranges from practically distilled to a saturated solution of various salts, from extremely clean to extremely polluted.

Interest in the life of inland waters arose from diverse directions: from the naturalist with a pond net, from the angler with a fishing rod, from the commercial fisherman with nets, lines, and traps, from those concerned at the growing pollution of rivers and streams, from the engineer responsible for supply of pure water, from the doctor concerned with water-related diseases. Parallel with such practical interests the pure scientist wanted to find out just how plants and animals were born, lived, and died below the surface of water, where observation was more difficult than on land. During the latter part of the nineteenth century and early in the twentieth these interests were developed in Britain by many institutions: by natural history societies, by the National Federation of Anglers, Salmon and Trout Association and Flyfisher's Club, by the several Institutions of engineers and by organizations for the medical fraternity. The relation between health and water was stimulated by the identification of water-borne pathogens, and around the turn of the century by discoveries which revealed the transmission of diseases such as malaria and schistosomiasis. However, there was no organization in Britain specifically for limnological scientists until much later.

On the European continent and also in North America progress was different. As early as 1868 Ernst Heinrich Haeckel had coined the word 'oecology' in his *Natural history of creation*, and in 1872 F. A. Forel, living in the village of Morges in Switzerland, settled down to a lifetime's investigation of the limnology of Lake Geneva. His main publications ran to three volumes, the first appearing in 1892. In the meantime in 1884 scientific investigations on inland fishery problems had started in Hungary with a view to legislation.

The first research station for fundamental limnology was established at Plön in Schleswig-Holstein in 1890 by Otto Zacharias. At first a private venture, this was soon supported by government and as a result Germany took the lead in developing both theoretical and applied limnology in

Europe until the First World War, and indeed, under the direction of August Thienemann, during the period between the wars. A number of other European limnological stations were soon established: Hillerød in Denmark was opened in 1900, Lunz-am-See in Austria in 1905, Aneboda in Sweden in 1908, and later Tihany on Lake Balaton in Hungary and an Italian institute on Lake Maggiore.

In the USA with its great variety of inland waters several stations were founded likewise during those early years, attached to universities. The first was by the University of Illinois in 1894, but the best known is that in Wisconsin where highly important contributions to theoretical studies were made by E. A. Burge and C. Juday in the early years of this century, and a tradition of advanced limnological research continues strongly to this day.

It is a little sad to record that Great Britain, which had led the world in so many branches of science, lagged behind in this one. The reason was that British aquatic biologists were drawn to the seas rather than to inland waters. A visiting authority from the United States, Charles A. Kofoid, who toured the research stations of Europe in 1908 and 1909, wrote: 'The direct support of biological stations by educational funds of local or state origin, often in connection with universities, so generally prevalent in other European countries, is almost wholly lacking in Great Britain.' But Kofoid also mentioned the 'relatively very large absorption of the funds and activities of the British stations in scientific fisheries work', which he described as 'unsurpassed in its excellence and effectiveness'.

Among the first scientists to do serious work on Britain's inland waters was Sir John Murray who had been with the *Challenger* expedition in 1873–6 and was later head of the *Challenger* office. In 1897–1909 he organized a survey of the Scottish freshwater lochs as a private venture. It was concerned primarily with sounding and plotting depth contours, but the theory of circulation of water in the deep lakes was elucidated from temperature measurements and collections were made of plankton. The contribution of this survey to the development of limnology has been underrated; some of the supplementary chapters are particularly illuminating. Then in 1901 in Norfolk the two brothers Gurney, Eustace and Robert, established a private research station on Sutton Broad and carried out vigorous hydrobiological research for a number of years.

At about the same time attention was drawn to British rivers by the Royal Commission on sewage disposal which was appointed in 1898. This was a remarkable venture, well ahead of its time internationally. Its final report was published in 1915 and the methods and standards have stood until quite recently. However, the mobility of water renders the study of rivers more difficult than of lakes and it was not until after the First

World War that major research on them was undertaken. Then the Ministry of Agriculture and Fisheries set about comparing conditions in uncontaminated and polluted reaches. Between the two wars the Ministry's staff conducted surveys of several rivers, particularly the River Tees, the primary object being to discover the effect of pollution on animals and plants. During the same period staff of the University College of Wales, stimulated by Kathleen Carpenter, investigated stream faunas and the effect on them of pollution from lead mines.

In northern England W. H. Pearsall, who had returned from the First World War deafened but with a tremendous enthusiasm for plant ecology, together with his schoolmaster father, concentrated on the Lake District. They did not pay too much attention to the classification of lakes which had emerged from the work at Plön in Germany and was based largely on Chironomid midges and oxygen content of the water. W. H. Pearsall recognized that from the time it is born every lake is destined to die from the deposition of silt and growth of vegetation, until it turns into swamp and ultimately into dry land. With this in mind he arranged the various lakes and tarns of the Lake District into a kind of evolutionary or development series. The series commenced with those of very pure water like Wastwater and Buttermere; it progressed with ever increasing productivity through intermediate stages such as Derwent Water, to Windermere which was already suffering a measure of pollution from sewage; and finally to the truly eutrophic lakes such as Esthwaite Water and Bassenthwaite. The end of the series was amply demonstrated by hollows which, like the lakes and tarns had been created by glaciers, but in which former open water had become clogged by swamp and accumulation of peat. He found that the flora, both the planktonic algae and rooted vegetation, had characteristic differences according to this progression. Similarly with the fauna, the fish in pure water lakes were dominated by char and trout, in the advanced more productive lakes by perch and pike, and in swamps Amphibia tended to dominate.

In working out this hypothesis Pearsall went into some detail and stressed, for example, the effect of the catchment area on the lake to which it drained: rocky at one end of the series, silted at the other. With hindsight the whole idea now seems obvious, but like other hypotheses which appear simple today—not excluding Darwin's natural selection—it has provided a pattern into which a vast amount of detailed research has been fitted subsequently.

University teaching in limnology was beginning at this time and J. T. Saunders had initiated the first course in the subject at Cambridge. I was a member of it and had made my first study in aquatic ecology—on the vertical movements of zoo-plankton—with Saunders at the little laboratory of Kastanienbaum on Lake Lucerne. During the 1920s Pearsall,

Saunders, and F. E. Fritch, the leading algologist who was professor at Queen Mary College in London, together with some others, were thinking about the establishment of a permanent freshwater biological laboratory in Britain. In 1927 Fritch gave his presidential address to Section K of the British Association for the Advancement of Science on this topic; a formative committee was established, and the Freshwater Biological Association of the British Empire came into being in 1929. The interest of universities, academic societies, fishermen, and water supply undertakings was aroused and, in spite of the slump, subscriptions towards establishing a centre totalled £575 in 1930, with the promise of a grant of about the same amount from government.

The time to start had arrived and after examining possibilities in the Norfolk Broads, the Cheshire meres, and other parts of Britain it was decided that Windermere was the most suitable location. Wray Castle, a gigantic 'medieval' folly which had been built in 1840 in beautiful grounds on the north-east shore of that lake, had recently come into the hands of the National Trust. It was partly occupied by the Youth Hostels Association so the FBA came in as tenant of the other part in 1931 and after a year took over the whole castle with its boathouses. The first research worker to arrive was Penelope Jenkin, another student of Saunders at Cambridge, and she was soon followed by two 'naturalists in residence', Philip Ullyott and Bobby Beauchamp, also from Cambridge. The facilities for research soon became known in the universities and the doors were opened to visiting workers.

The subscription list and the government grant increased steadily but slowly. In 1936 the Development Commission, as the responsible department of Government, sent a visiting group headed by Professor Graham Kerr of Glasgow to inspect the FBA and they recommended an increase in the grant sufficient to appoint a paid director and an assistant secretary. Thus by 1937 there was a modest salary available for me to take over from Pearsall and P. A. Buxton, who had been honorary director and honorary secretary of the Association respectively.

In 1937 there were four young research workers in residence in addition to myself, providing a team conducting research on major topics which contribute to the ecology of inland waters. T. T. Macan was concerned with invertebrate animals, especially insects. Before long he became deputy director, a post which he filled with distinction for some 35 years, by the end of which the Association's scientific staff had increased more than tenfold from that original four. C. H. Mortimer focused his work on the chemistry and physics of the aquatic environment: some years after the Second World War he moved to Scotland as director of the Scottish Marine Biological Association and later continued his distinguished career in the United States as a professor in Wisconsin. K. R. Allen, the fish biologist, was the first to leave Wray Castle: he married Rosa Bullen, who had been appointed assistant-secretary and came from New Zealand. Allen took a post there and conducted classical research on trout streams: later he proceeded to various senior posts including the directorship of the Pacific biological station of the fisheries research board of Canada at Nanaimo and chairman of the scientific committee of the International Whaling Commission. The botanical side was looked after by an attractive young lady from Austria, Marie Rosenberg, who concentrated on the algae. There were also, most of the time, several visiting research workers working on their own projects. This nucleus was supported by a group of laboratory assistants headed by George Thompson, a local boy who spent his whole life leading that group; it increased in numbers and abilities steadily from year to year.

Soon the scientific staff was augmented. Winifred Frost joined us from Ireland and spent the rest of her life on fish studies with the FBA. The algologist, J. W. G. Lund, came to Wray Castle after several years' experience in forensic laboratories at Birmingham; he soon established an international reputation. C. B. Taylor, initiated studies of freshwater bacteria financed by the Department of Scientific and Industrial Research. P. H. T. Hartley investigated course fish, financed by the National Federation of Anglers; he also studied birds and later took holy orders while retaining an expert interest in natural history.

During the war years some staff were drawn away for special work and others joined the group. Notable among the latter were E. D. Le Cren and Rosemary Lowe who were 'directed' to work at Wray Castle on wartime fisheries. Le Cren never left the FBA and is now its Director, Rosemary Lowe (McConnell) later did pioneering work on African lakes (see p. 26) and became a leading authority on tropical fish and fisheries. Noel Hynes came from London to do his research for a Ph.D. at Wray Castle on aquatic entomology; he later became a leading specialist on river ecology and water pollution and proceeded to a professorship in Canada.

All the staff and visitors, except those who got married, lived as well as worked in Wray Castle, so there was a social side to the establishment as well as laboratories and library. Inevitably colourful events occurred, many of them contrived by long-term visitors, in the early years Arthur Ramsay and later Gerald Swynnerton, both of whom had a penchant for practical jokes. During Swynnerton's period things frequently went bump in the night and it was dangerous to open almost any door for fear of being doused by a bucket of water. A feature of those years was that everyone worked extremely hard but kept no formal hours. A piece of research might involve 24 hours of continuous observation, after which the research worker might spend a day climbing a mountain or going fishing.

There is much to be said for a small group of research workers who, by informal use of time, can produce really exciting results. As the number of scientists and technical assistants increases the 'nine to five' mentality is difficult to avoid and some of the inspiration is lost. Our small team covered the major subjects involved in inland water ecology. We were able to search more deeply in the relatively confined environment of Windermere than had been possible in the vast extent and the much greater complexity of the African lakes.

Our researches were admirably supported by the honorary officers of the Association. The President was Reginald Beddington who was also President of the National Association of Fishery Boards and so in a position to make many valuable contacts. F. E. Fritsch was Chairman of Council, and among other founding fathers who continued to take a close personal interest, visiting Wray Castle on occasions, were W. H. Pearsall, J. T. Saunders, and P. A. Buxton who, as former honorary secretary of the Association, brought in the concepts of medical entomology and water-related diseases. Our honorary treasurer was Sir Albert Atkey of Nottingham who was also at the time President of the British Waterworks Association so we had ready contact with water engineers. Atkey, as a self-made man of business, also had great sympathy with our financial problems. At one time the Association had no funds with which to purchase a serviceable typewriter, but he secured a good one second hand. It had no road transport so he produced a large Austin saloon from his own garage in Nottingham at a remarkably low price. This vehicle served well throughout the war, not only for official journeys but sometimes on hire to the staff to get away from the isolation of Wray Castle to the bright lights (or blackout) of Ambleside or Bowness.

An early operation which was to provide background for many others was a detailed survey of the underwater configuration of Windermere. This was conducted by a survey officer loaned from the Admiralty, and he took some hundreds of traverses back and forth across the lake with the result that soon more was known about the depth contours and volumes of water for Windermere than probably for any other water of comparable size in the world. This was one of the early surveys to be done by echo-sounder which recorded continuously the depth of water over which the boat passed. When these recordings were examined in the laboratory it was apparent that, whereas in some places the bottom had produced clear abrupt echoes of the sound emitted at the surface, in other places the record was blurred and soft. It was a simple matter to check that the clear record was over rocky bottom and the blurred record over mud. With a little experience it was possible to distinguish also sandy bottom, areas with rooted vegetation, and in some cases shoals of fish. This was during the very early days of the application of echo-sounding so that these discoveries were exciting.

Even more important was the way in which hard rock on the floor of the lake, when covered with layers of soft deposits, could be discerned so that in some parts of the lake there was opportunity to estimate the depth of deposit which had accumulated since the original gouging out of the Windermere basin by a glacier during the ice age. Very soon Clifford Mortimer, who was particularly involved in these investigations, was devising various instruments in order to raise cores of bottom deposit for examination in the laboratory. Even a simple drainpipe rammed down into the mud indicated what treasures there were in store for future palaeolimnologists. Later, in co-operation with an engineer, B. M. Jenkin, father of Penelope Jenkin, much more effective devices were designed and constructed in order to obtain cores of the bottom deposits with the minimum disturbance of their content. About this time a young lady student from Reading University, Winifred Pennington, came to live and work at the Castle in order to study geology and botany. We put her on to examining these cores, and so started the British tradition of palaeolimnology which has produced exciting results not only from many British lakes but from other parts of the world. Miss Pennington (Mrs Tutin as she became) continues to be a leading exponent of this subject, now with FRS after her name.

These cores gave promise of factual evidence of Pearsall's hypothesis, mentioned earlier, that the lakes which occupy glacier cut and moraine-dammed depressions, represent a series of stages through which every such lake has passed, or will pass, during the course of its history. This indeed was worked out in detail by Winifred Pennington in subsequent years and, in association with Pearsall, she related her results to archaeological and historical evidence. As affecting Windermere's biology the most important change was revealed by the phytoplankton which had left its record in the deposits. The top 20 cm of the cores taken from the deep undisturbed part of Windermere were found to contain abundant remains of the diatom *Asterionella* which is dominant in the spring plankton today but was rare in the earlier deposits. It seemed probable that the factor causing this change was the development of human settlements on the lake shore and especially the introduction of water-borne sewage. By applying methods of dating the deposits through the rate of sedimentation of mineral matter, and in some cores by varved clays, it was concluded that this change took place certainly not more than 220 years ago and may have been considerably more recent. The change may therefore have started in the eighteenth century but there can be little doubt that it was accelerated when the railway reached Windermere in 1847. Not only did the railway bring an influx of tourists to the lakeside villages of Bowness and Ambleside, but it allowed the creation of the town of Windermere and building of many residences round the shores. Water-borne sewage from all these habitations has been poured into the lake with or without treatment ever since.

The process of unravelling the aquatic ecology involved not only Windermere itself but the whole catchment area, its climate, and its geology which controlled the quantity and quality of water, its movements within the lake, layering in summer and turnover in winter. Measurements of inflowing and outflowing water were essential, and it is somewhat astonishing to recall, considering the importance of water and river flow in the national economy today, that at that time, only 40 years ago, there was no kind of official system for recording river and stream flows. For advice we turned to Captain Maclean who had created the private organization of 'River Flow Records', and only with his assistance was it possible to put together the outlines of a water budget for the Windermere catchment. This work involved the installation of a good many additional rain gauges including some recording gauges near mountain tops, as well as water level gauges, current meters, and the construction of stage-flow curves.

All this concentration of the FBA's work on Windermere carried on the tradition of pure scientific research for which the venture had been initiated, each member of the team doing his or her own research but in a co-ordinated programme. However, my own ideas were somewhat broader. It seemed to me, and I was backed up in this by most of the Association's Council, that it was important to initiate research programmes elsewhere, in other waters of the Lake District, in other parts of Britain, and, when opportunities offered, in overseas parts of the then Empire. Even if expansion was to cause the depth of research in the particular situation of Windermere to be sacrificed somewhat for the width of a broader view, more support for the Association's activities would be forthcoming and this was important. Thus in the Association's report for the year ended 1939 one reads that research on several rivers in the south of England had come under the scientific guidance of the FBA by an arrangement with the then active University College, Southampton, Branch for Southern Rivers which was under the stimulating leadership of John Berry. In addition a centre in East Anglia had been established for Peter Hartley's coarse fish research. Nearer at home a new investigation, calculated to be of interest especially to the water industry, had been initiated on the extremely pure waters of Thirlmere and on Haweswater before its impoundment to augment Manchester's supply. The new unit of bacteriology, in association with algology and chemistry, had started a study on Esthwaite Water for comparison with the work on Windermere. In overseas countries at that time Beauchamp, the former naturalist in charge at Wray

Castle, was working on Lake Tanganyika, Allen, who had already extended the FBA's interest to salmon in Caithress and the River Eden, was on New Zealand streams, and Swynnerton had joined the game department in Tanganyika Territory. Another initiative in 1939 was to start a series of '*Scientific Publications*' of the Association consisting of booklets most of which provide ready means of identification of the British aquatic flora and fauna. This series has continued and at 1982 numbered 46 together with a good many second and third editions.

The FBA, like its predecessor and counterpart the Marine Biological Association based at Plymouth, has always welcomed scientific visitors for short or long periods and indeed has benefited greatly from association with them, not only in the scientific results which they produced, but in getting the work more widely known. Even in those pre-war days there was a steady flow of visitors and each year all members of staff contributed to a fortnight's intensive teaching at the 'Easter Class' which consisted of students drawn from a number of universities.

A few weeks before the war broke out in August 1939 we received at Wray Castle one of our most colourful visitors, Julian Rzoska, who at that time was a professor in Poland and came to study the zooplankton of Windermere. He entertained us with his command of language and antics on the croquet field and then, as war appeared inevitable, bolted back to Poland to join the armed forces. Apart from his war-time experiences and escape to Britain which make an epic in themselves, the FBA has never lost touch with Julian Rzoska. After the war years, being a stateless person, initially with no means of communication with his family still in Poland, he built up a remarkable investigation unit on the River Nile at Khartoum University. Later he initiated many students into freshwater biology during a period of teaching in London, and for ten years was the scientific co-ordinator of the freshwater productivity section of the International Biological Programme.

Inevitably the war years caused some disruption in the activities based on Wray Castle. The scientific staff, being in reserved occupation, were instructed to stay put until required for specific war purposes, but gradually they became dispersed, Macan to malaria control among troops in the Far East, Hynes to locust control in East Africa, myself to the Middle East. But there remained a nucleus which was added to by some staff from the British Museum of Natural History with their national collections stored in the comparative safety of Wray Castle, and from the government's fisheries station at Lowestoft.

These additional scientists played a full part in the wartime work. W. E. China from the Natural History Museum was my sergeant in the local platoon of the Home Guard and occupied the director's chair during periods when I was called to the Middle East; he later proceeded to the

Keepership of Entomology at South Kensington. Kimmins extended greatly our knowledge of caddis flies and several other groups of aquatic insects. Buchanan-Wollaston from Lowestoft added a new dimension to our instrumentation for research, and his wife took charge of the housekeeping and problems of wartime catering for the diverse group who lived in or worked at the castle. During the war supplies of marine fish were reduced almost to nothing so our group at Wray Castle focused attention on obtaining food fish from inland waters. Our pre-war contacts with scientific colleagues in Central Europe had informed us that nearly every water there produced fish, with cropping rates ranging from about 5 to 50 kg per hectare, and much more than that in the well-developed fish farming industry where carp was the main product. In central Europe, with no marine resources, there was little or no prejudice against eating the so-called 'coarse fish', some of which were regarded as luxuries, whereas in Britain hardly any were eaten except by the Jewish community. The war years did not seem likely to provide either the time or facilities for developing a fish-farming industry comparable with that on the continent, but the natural productivity of lakes, reservoirs, and rivers could surely produce some worth-while crops. The lakes and rivers of England and Scotland comprise some 150 000 hectares, and those of Ireland about the same. As a contribution to British food two species of fish seemed to offer the best opportunities—eels and perch. For eels there was already a market, much of which had formerly been supplied from the Continent; for perch it might be possible to create one, for their flavour is excellent though the fillets which can be cut from the scaly and bony body are rather small.

The eel has two phases: the yellow eel which grows in fresh water for up to ten or fifteen years after the tiny elvers have crossed the Atlantic from the breeding grounds in the neighbourhood of the Sargasso sea and have ascended our rivers and streams; and the silver eel into which the yellow eel transforms itself when it gets the urge to cross the Atlantic once again to the breeding grounds. In this transformation the eel develops its swimming muscles, lays down quantities of fat as an energy reserve, and enlarges its eyes. It then stops feeding and allows its digestive system to degenerate. The colour darkens and a silvery sheen develops. Then the eels set off downstream and across the Atlantic to the Sargasso Sea where all the silver eels from Europe (and Eastern North America) congregate to breed and then die.

The objective of eel fisheries clearly should be to catch the eels in the silver stage when they are more nutritious than in the yellow stage, and when they gather in shoals for the downstream migration to the sea. They choose dark stormy moonless nights in the autumn, when rivers are in flood, for this purpose. In the old days silver eels used to be trapped all over Britain at favourable spots where there was a weir at the outlet of a lake, or at a mill. A small canal was constructed at the side of the weir and the water arranged to fall through a grid where the eels were stranded. They wriggled sideways into a storage box filled with water from which they could be scooped, sometimes by the hundredweight, packed in boxes with wet moss, and sent alive to market. Eels, unlike most fishes, can live quite happily in the air provided the environment is damp. Sometimes, in their urge to reach the sea, silver eels will wriggle across land if there is no other way.

Unfortunately most of the old traps for silver eels in Britain had not been operated for years and had fallen into disrepair or had completely disintegrated. Presumably the market value of eels had fallen so low compared with marine fish that the extra work involved by the mill owner was not worth while. But during wartime the market value was well up so, in consultation with the Ministry of Agriculture and Fisheries, we set about encouraging the reconstruction of traps. There had been a number of eel traps in the Lake District and one or two were still in operation, notably one at Newby Bridge by the weir which controls the outfall of Windermere to the River Leven. None of these traps were particularly efficient because, on the flood-ridden nights selected by the eels for migration, there was so much water pouring over the weirs that only a small proportion of the migrating eels would take the channel leading to the trap. Therefore our research was focused on the behaviour of eels and the factors which induced them to turn one way or another in their downstream migration.

Any form of mechanical grid to direct the fish to the trap entrance proved of little use because the flooded streams brought down masses of leaves and other debris. We tried electrifying widely spaced grills with the idea of deflecting the eels while letting through the debris, but it seemed that the rushing water and the speed of travel of the fish was too great for this to have much effect. The most promising line therefore was to work on the well-established aversion of silver eels to migrate on moonlit nights and to divert the eels towards the trap entrance with the aid of artificial moonlight. This was not easy, particularly with the blackout regulations of wartime, but many were the gadgets created in our workshop to produce strings of artificial lights underwater, shaded in such a way that they were obvious only to oncoming eels. Eventually a system was adopted, some half-dozen eel traps in the Lake District were got into working order, rigged with our equipment and manned on favourable nights in the worst autumn weather by volunteer students. Sometimes during a midnight storm the branch of a tree would come floating down and get entangled in the equipment and someone would have to strip off and clear it. Such are the delights of ecological research. Most of this work had to be conducted in the field, but we also constructed an artificial stream in the cellars of Wray Castle where the reactions of individual eels could be tested.

Some of these experiments in the field and in the laboratory definitely had the effect of diverting eels, but it cannot be claimed that the results were as favourable as we had originally hoped, or that our various improvements to the traditional system of eel trapping were adopted commercially. However, there is no doubt that eel fisheries were much expanded during the war, and even from our own experimental installations we sent a good many consignments to Billingsgate fish market.

Our other fisheries exercise during the war, concerned with perch, had more lasting results. Not only did it produce a considerable quantity of food but also initiated a large-scale experiment in the population dynamics of fish which was subsequently followed through by the FBA staff for more than three decades, and still goes on. In the 1930s and early 1940s the fish populations of Windermere were characterized by a prodigious number of perch, nearly all of very small size weighing no more than an ounce or two each. Tourists could enjoy catching them by the dozen almost anywhere in the shallower waters but they were of no value for any other purpose except as food for cormorants and pike. There were trout as well, which produced some fly-fishing in the spring and some big specimens taken on the troll. There were also in the deep waters fair numbers of the famous Windermere char (*Salvelinus willughbii*) which were taken only by semi-professional local fishermen who rowed slowly about the lake, two rods set to port and starboard, each with a bell at its tip and a long line to a heavy weight with streamers set at intervals and baited with spinners. The best baits allegedly were beaten out of golden sovereigns. Char was a special delicacy at local hotels and potted Windermere char was occasionally available at Fortnum and Masons in London.

Early records of the fishery of Windermere show quite clearly that an increase of perch and pike and reduction of trout and char was progressive from about 1860 to 1940. By 1869 the Board of Conservators, which had been established in 1866, approached the Home Office about a closed season and larger mesh of nets used for catching trout and char. In 1879, when a closed season was imposed, although not a larger mesh for nets, a full enquiry into the freshwater fisheries of the Lake District showed that complaints about deterioration were widespread. During the First World War netting on Windermere was increased in connection with food shortage, but the nets were finally taken off in 1922. Since then, however, the fishery continued to deteriorate. Authors who have made reference to this subject in subsequent years all agreed that the perch had become a curse in Windermere, as also in some other lakes. As their numbers

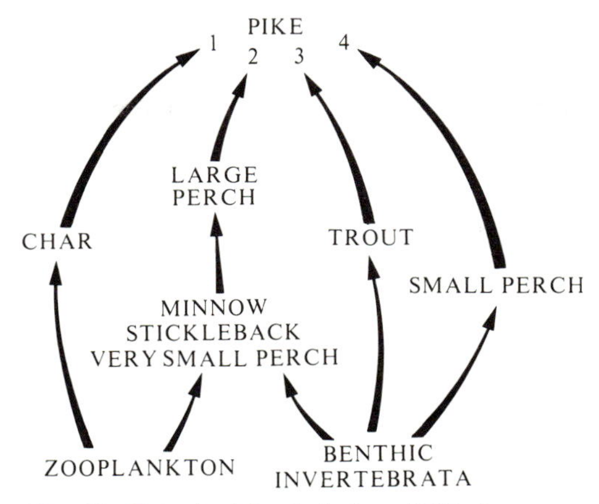

Fig. 10. Principal food-chains of Windermere.

increased, they had become so dwarfed in size as to be practically valueless, either to the angler or the housewife. There was reason to believe, moreover, that the change in fish ecology, which is illustrated by the principal food chains of Windermere (Fig. 10), was associated with an overall increase in biological productivity stimulated by the additions of sewage. It seemed to our group of research workers at Wray Castle quite impossible to reverse the process by cutting off the sewage, for even with full treatment the nutritive salts which it contained would enter the lake; but perhaps the fishery might be brought back to something of its earlier value by vigorous management, starting with a reduction in the perch population.

Four questions presented themselves: (1) Could the superabundance of small perch be converted somehow into edible food during wartime? (2) Would a drastic reduction in the perch population be followed by an increase in the growth-rate of the survivors so that in post-war years they would provide better angling? (3) Would the trout and char, which competed with the perch for their food, increase in numbers when perch were reduced? and (4) If so what about the pike which at present depended mainly on perch? At the back of our minds also were the much larger problems of fisheries management in the seas surrounding Britain where it was exceedingly difficult to obtain accurate estimates of populations and of the influences of overfishing on growth-rates, competition between species and predation. The theoretical considerations behind such questions might be clarified by an experiment in the confined waters of Windermere with its relatively few species of fish whose relations to each other could be studied intensively.

The perch of Windermere had already been the subject of research by

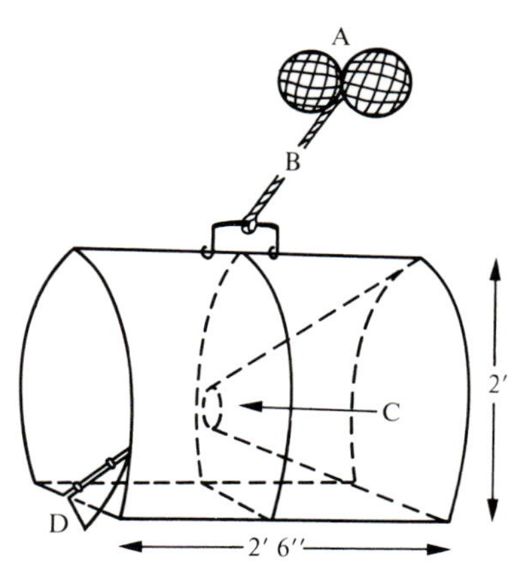

Fig. 11. Perch trap as designed for Windermere and used later in many lakes and reservoirs. A glass floats in string bags; B rope 7–9 m long; C funnel with opening 16 mm diameter, D. flap door.

K. R. Allen who had determined their feeding habits and had demonstrated that they undergo a seasonal migration, spending the summer months in the shallow water after their spawning in May and moving to the bottom in depths of 20–25 metres during the winter. The first problem therefore was to catch the perch in maximum numbers at minimum expense. Seine netting the shallows in summer was too labour-intensive; we thought of trawling the deeper water in winter but that would be very difficult on the soft muddy bottom with rocky outcrops; so I harked back to the wonderful variety of basketwork fish traps used by the tribes around Lake Victoria, and we constructed some experimental ones of wire-netting on a framework of stout fencing wire with a re-entrant mouth, each with a float at the surface like a lobster-pot. We set a series of these at prescribed depths from near the shore to 30 metres, emptied them every day to weigh and count the catch and thereby determined the time of year and precise depth for maximum catches. This proved to be in the spring from the end of April for about six weeks, at a depth of 8 to 3 m as the fish accumulated to spawn after migrating in shore from deeper water. In those circumstances the catches per trap were quite phenomenal, many hundreds weighing many kilogrammes each day. These trials showed that a few hundred traps fished around the lake shore at the right time of year and the right depth could catch perch by the ton with very little labour.

The next problem was how to utilize the fish when landed. They proved to be good pig food, and that was how many of the experimental catches were disposed of, but what about human food? At this point help came from a leading personality in the north sea fishery which had practically closed down during the war. He was Mr F. Parkes, who owned trawlers and fish canneries, and he was impressed with the possibility of canning the perch like sardines. In size they were comparable; the pressure-cooking involved in canning would soften the perch's troublesome bones and scales; and we thought of the selling name 'perchines'. Parkes arranged for trial lots in various media and we voted for tomato sauce.

We arranged a meeting of local fishermen and asked for their co-operation in looking after 50 or so perch traps each along prescribed sections of the lake shore. In the event some 30 fishermen and fisherwomen did this regularly for nearly two months each year from 1941 to 1947. They treated it as voluntary war work and were interested in its possible beneficial effect on their beloved char and trout fishing. The first season, 1941, was limited to the north basin of Windermere and produced 25½ tonnes, which, at average weight of little more than 25 g, meant an awful lot of perch. They were all packed at the lakeside in boxes, collected by lorry, taken to British Fish Canners of Leeds and turned into perchines. Some went to the fighting services overseas, and a year or so later, when working in the Middle East, I heard some reactions to them—not all complimentary!

It was entirely predictable that the initial rate of fishing would not be maintained. Thus in 1942 each trap in the north basin, fished for the second year running, caught only about half as much as the year before. The number of traps fished was nearly doubled, however, to over 600,

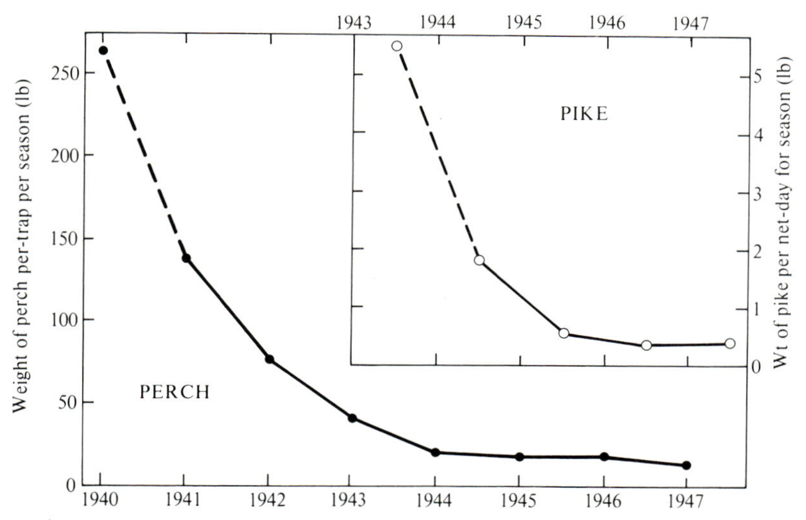

Fig. 12. Experiment with populations of fish in Windermere: catches of perch and pike during years 1940–7.

and the south basin of Windermere was included as well as the north, so the total catch was increased to over 30 tonnes. A similar rate of reduction in catch per trap was recorded in 1943 and again in 1944, by which time we reckoned that the original objective of a drastic reduction in the perch population had been achieved. Although in subsequent years the canning operations were discontinued the perch trapping was carried on experimentally and the catch per trap remained remarkably steady for the next 10 years at approximately one tenth of what it was in 1941. This provided a classic example of overfishing, reducing a virgin population rapidly to a rock-bottom sustained yield. Four years' intensive fishing with traps for perch in Windermere had produced just about the same result as 25 years' fishing with gill nets for *Tilapia* in the Kavirondo Gulf of Lake Victoria (see Chapter 2).

Each year the perch traps caught a few hundredweight of pike, some eels and an occasional trout, though never a char. We were anxious about the effect of the pike on the meagre populations of trout and char since the pike's normal diet of perch had been removed, so the next stage of fishery management on Windermere was introduced in 1944. This was the netting of pike with gill nets of 13 cm mesh (6½ cm knot to knot) which would allow all char and all trout except a few very large specimens to pass through undamaged. After experimental trials in 1943 the succeeding three seasons of pike netting accounted for 1472 pike weighing 5500 kg. They were all examined by Winifred Frost who found among other things that the sexes were almost equal in numbers, but females averaged about 5 kg compared with 2½ kg for males. The effect of this fishery on the pike was to change the balance of the pike population from one dominated by old large pike to an abundance of young and small pike, few of whom were able to survive to more than five or six years of age because of the size-selective mortality from the fishery. Winifred Frost and Charlotte Kipling were able to find out a great deal about the biology of the pike from the detailed examination of all the specimens caught by the nets and from marking experiments. During this period the nets also caught 289 trout averaging 2 kg. These big trout were predators on smaller fish and so, like the pike, were best out of the way. Regular sample netting of char as they came from the depths into the shallow-water breeding grounds was also instituted at this time in order to obtain an index of char population.

Following the first successful season of perch trapping in Windermere the method was extended to other lakes in the Lake District, and also to certain lochs in Scotland which were known to be suffering likewise from a surfeit of perch. In some cases these supplementary perch fisheries were able to market the product and add to wartime food supplies, and in most cases perch trapping was found to be a useful method of reducing coarse fish in waters, such as Loch Leven, where the prime objective is fly-fishing for trout.

In the history of the FBA, which now runs to half a century, as celebrated in 1979, the wartime contribution of food fished from Windermere and other lakes is a mere incident. However, the change in the balance of fish populations between the four important species, char, trout, perch, and pike, and the effect of this on the ecology of Windermere has been followed through continuously until today. Perch trapping by the FBA still continues on an experimental basis and a great deal of information about the fish populations has been published by E.D. Le Cren who took a prominent part in the original experiment and is now Director of the Association, and by other staff. Soon the perch began to grow more rapidly, so the average weight of those caught by anglers became much larger than previously. If the opinion of anglers is anything to go by there are today more trout, and it seems quite definite that the char are more numerous. The pike continue to be controlled by netting and it seems that a new balance between the populations had been struck. However, in 1976 the perch population suffered another quite unexpected and devastating setback in the form of a disease, specific to the species, which initially appeared to have practically wiped them out of Windermere. The perch population now seems to be recovering, so here is another opportunity to watch a new balance of fish populations re-establish itself.

In this account of the experiment with fish populations the focus has been on man-management, but nature has also played its part. Indeed a striking finding of the whole study has been variability from year to year of brood success of perch, pike, and char. It seems likely that variations in climate, which affect life in water almost as much as on land but in different ways, is the primary cause of this. In many situations the natural changes in population balance are as important as those caused by human interference.

The period in the Lake District covered by this chapter, with a small team conducting research in a relatively small locality, was perhaps typical of the general state of ecological science when nearing the middle of the century. Ours was by no means the only such group working intensively on ecological problems in different environments in many parts of the world. Out of these studies emerged a number of principles which became of growing importance in later years in relation to the development and utilization of natural resources. Among these principles were those of production biology; control of the quantity and quality of vegetation by chemical nutrients and availability of light; the influence of limiting factors; transfer of energy through intricate food relations of different species and populations, resulting in a complex food web rather than a simple food chain; the importance of past biological and physical events which led to present situations and help in predicting the future. Also our work enforced a principle which is of high importance in all forms of land and water planning and development, namely, consideration of the catchment area as a whole rather than focusing attention on particular parts of it. Research on a lake such as Windermere can become so absorbing that one can almost forget that it has inlets and outlet, but many of the ecological problems can be solved only when the total catchment area is taken into account—the rainfall, its quality as well as quantity, the geology, soils, vegetation, streams, marshes, and tarns in the upper catchment, and the outflowing river which introduces migratory fish from the sea. Added to these natural factors are the influences of man—agriculture, animal industry, forestry, fishing, and the disposal of waste. All these add up to an ecological complex of which the lake itself is only a part.

After the war scientists and laboratory staff who had been away on war duties returned to the fold and plans were made for a considerable expansion of the FBA in terms of staff and visitors. This led to the conclusion that Wray Castle, large as it is but restricted in how its space could be modified for scientific purposes, was unsuitable as the FBA's permanent headquarters. The Association's Council, in consultation with the Government from which an ever-increasing share of the funds was provided, decided that the headquarters should continue to be based on Windermere, so all large houses around the lake which were or might come on the market were examined and assessed for the purpose. In the event the Ferry Hotel facing Bowness was selected and with many adaptations and substantial additions has become an excellent headquarters for the greatly enlarged organization. But that happened after I had been called to other work on behalf of overseas countries, so continuation of the story needs to be written by others. This indeed has already been done in the series of annual reports and in summary by G. F. Fogg (1979) when the Association reached its half century.

5
The Middle East (1943–1946)

The first half of this century, and particularly the periods following the two great wars, has been notable for a mass of literature about Middle Eastern politics, recording events, ideas, and opinions which did much to create the situations of today. This chapter is not much concerned with politics, however, but it attempts to summarize the level of scientific, and especially ecological, understanding which came to be recognized after the Second World War as a basis for economic and social development. Up to that time this knowledge, in terms of Western culture, had come in most countries more from alien than from indigenous interests, especially from British, French, and Italian men and women, who, as long-term residents or as short-term visitors, had amassed information.

Most of the interest had centred around archaeology and history since some of the countries of the Middle East had been the cradle of great civilizations. The science of hydrology, which had been initiated by ancient Egyptians on the Nile, was also making progress since much of the region depended on irrigation for its supplies of food and fibre. But the main productive and applied sciences, concerned with the soil, domestic animals, fisheries, health, and mineral resources, had lagged some way behind, with a few notable exceptions such as cotton-growing and oil exploration. Education was mostly of a traditional character, much of it based on the Koran, but modern universities were sprouting in nearly every country so new generations of graduates were taking a new look at their own towns and countryside. Thus foundations were being laid for the technological and social advances which came rapidly after the Second World War. Foundations were being laid also for the environmental revolution which, however, cannot be said as yet to have had marked effect on all the countries of the Middle East.

What countries are included in this region? The 'Near East', 'Middle East', and 'Far East' have always been unsatisfactory names for geographical regions because the definition of their boundaries has been so varied. The situation had become so confusing that Sir George Clerk tried to sort it out in his Presidential address to the Royal Geographical Society in 1944, by suggesting that 'no part of Africa except Egypt belongs to the Near or Middle East'. He drew the dividing line between Near and Middle East immediately to the east of Syria, Lebanon, Palestine, and Egypt. This however was no help during the war, when the Middle East

Fig. 13. Sketch map of the Middle East at the time of the Second World War.

Command and the Middle East Supply Centre based in Cairo had accepted quite a different grouping of countries. Therefore, while recognizing its shortcomings, the term 'Middle East' is used here for that group of countries for which the Middle East Supply Centre (MESC) had some responsibility. There was in fact an inner and an outer group. The inner group, with the focus at Cairo as the nodal point for communications, consisted of the main Arab block, namely Egypt, the northern Sudan, Palestine, Transjordan, Lebanon, Syria, and Iraq, with Cyprus as an island addition. The outer group, proceeding clockwise, included Persia, Saudi Arabia, Yemen and Sheikdoms of the Persian Gulf, Aden colony and Protectorate, Ethiopia, Eritrea, the southern Sudan, Cyranaica, Tripolitania, and Malta. Turkey was also included for some purposes, and occasionally also the Somalilands.

MESC was established by the British Government in 1941 but became a joint Anglo-American agency in 1942. Attached to GHQ in Cairo its purpose was to ensure that the civil populations in the Middle East countries were provided with those supplies which were essential for their livelihood in wartime. It was necessary to see that imports were kept as low as possible in view of limitations in shipping capacity and land transportation, so that as much tonnage as possible could be devoted to the vast supplies of war materials needed for the desert campaign in North Africa, and for aid to Russia through the Persian Gulf. In the event, apart from war materials, imports into the principal Middle East countries were reduced from over 6 million tonnes a year in pre-war years to about 1½ million tonnes in 1944.

MESC dealt with every sort of supply, even aphrodisiacs for very important persons, but food was obviously one of the most essential and it was quickly apparent that increased agricultural production and change in some of the cropping systems, with controlled importation and distribution of fertilizers and machinery, could provide one of the greatest savings in tonnage for civilian needs. Not only agriculture was involved in this but also the whole pattern of natural resources, land and water use, and also the extremely diverse social and administrative organization in the region as a whole.

In 1943, when the North African campaign was nearing its end, it was felt that an objective study of the overall situation as it was, which pinpointed the needs particularly in scientific knowledge, could provide a valuable basis not only for the wartime purposes of MESC but also for post-war economic and social reconstruction in which British and American co-operation might well have a hand. Accordingly the Director General of MESC who was Robert Jackson, widely known to his friends as Jacko, and the head of its food and agricultural division, Keith Murray, with the help of Leonard Elmhurst of Dartington Hall, Lord Hurcomb, and Max Nicholson of the Ministry of Shipping in London, appointed a scientific advisory mission to the Centre.

Initially this mission consisted of two, Bernard Keen, who had been for some years deputy director to Sir John Russell at Rothamsted Experimental Station for the soil sciences, and myself, then director of the Freshwater Biological Association. Keen was to cover the agricultural subjects; I had the rest which included surveying, geology, meteorology, water, wild plants and animals, forestry, fisheries, human health, population, and social studies. This was a very wide field, but was not entirely new to me after experience on the African Survey. The reports of these two, each of which was later published in book form (Worthington 1946; Keen 1946) became the main permanent record of the mission. Later however the mission was added to by a specialist in animal industry and an educationist, in the persons of Norman Wright, who was at that time director of the Hannah Dairy Research Institute in Scotland, and Dr H. B. Allen from the American University at Beirut in the Lebanon. Allen's education report was also published (1946), but unfortunately Wright's never reached that stage: he was soon appointed to an absorbing position in the Ministry of Agriculture, Fisheries, and Food in London as Chief Scientist in succession to Sir Jack Drummond, who, with his wife and daughter, had been tragically murdered in France while on a camping holiday.

Keen and I were due in Cairo to start this study in November 1943, but travel from London to Cairo at that time was not what it is today. The Royal Air Force was the only means and my journey took six days, two of which were taken up with delays before starting owing to the weather and tactical activities by various airforces. Once off the water in a Sunderland flying-boat and after a first stop at Foynes in Ireland, we took a wide detour into the Atlantic in order to keep as far as might be from patrolling German aircraft, before coming into Lisbon. Then other stops at Gibraltar and the Island of Jerba in the Gulf of Gabes before the last leg over battlegrounds of the Western Desert, and finally cutting the waters of the Nile at Cairo.

Bernard Keen arrived within a day or so and we made contact with colleagues in MESC which had staff in every country of the region. We also made a date with Jacko to discuss how to organize the mission: there were no written terms of reference, so, after a short period of getting to know each other, he whistled up his secretary and in characteristic form dictated: 'Begin—the Scientific Advisors will travel as widely as possible throughout the Middle East and will report; they will be provided with all the facilities which they require. End.' This, when passed to the several departments of MESC, was good enough for us so we made a plan of operations. In the event it consisted of a spell of field-work in the Middle East, from November 1943 to May 1944 and then, after a period of consultations in England, another spell in the field from October 1944 to March 1945 followed in England by sorting out masses of notes, writing and discussing reports, and finally seeing them through the press.

Nearly every country in the region was visited at least once and most of them twice. Travel from one to another was perforce mostly by air, but extensive tours were made on the ground through the assistance of the local governments, the military authorities, and representatives of MESC. A good number of the key figures in military administrations, which by then were in charge of the occupied Italian colonies of Ethiopia, Eritrea, Cyrenaica, and Tripolitania, as well as officials in other countries, turned out to be old acquaintances whom I had known during the African Survey.

Air travel was entirely via the RAF or its American counterpart and, since passenger space was strictly limited, whether one achieved the lift or not often depended on one's titular rank. The scientific advisers were given high priority, but sometimes it was not sufficient. On one memorable occasion, being on a mission from Cairo to Addis Ababa, I got to Tobruk in order to pick up a plane on its way from London. At the time my priority was that of Brigadier and when the plane landed I was informed that my seat was already occupied by a General. He was not around so, assuming he had remained on the plane for its brief fuelling stop, I watched the plane take off for Addis with some chagrin. At the office I enquired who the important person was. The Corporal in charge referred to the list and said 'Oh dear, there has been a mistake; it was a consignment of chairs addressed to Toc H in Addis Ababa with the priority of Lieutenant'. Some joker had added the word 'General' after 'Lieutenant' on the manifest. Later on, in company with Keen and Allen, I reached Addis by way of a three-day bus ride from Asmara through that wonderful country of mountains and plateaux. After a week in Ethiopia, which included inspecting the origin of the Blue Nile flowing from Lake Tsana, there followed an exciting safari with a military convoy southwards to Nairobi, camping all the way. We passed the series of small rift valley lakes in Ethiopia, then through the high mountain forests with many colobus monkeys and a superb sighting of a black leopard; down the escarpment into Northern Kenya, across the hot black lava desert to Marsabit, now a National Park; and so, passing between Mount Kenya and the Aberdare Range, to the civilization of Kenya's capital.

Although at this time the military campaigns in North Africa were finished and the scene of conflict had moved to Italy, there were still hazards in Middle East travel. On one occasion, as passenger in a small plane from Cairo to Tripoli with a young American pilot, the weather closed in and there were only occasional glimpses of the ground. The pilot turned to me and said 'Have you by any chance a map, I have forgotten to bring mine?' I had with me nothing better than a pocket atlas. Luckily we got a brief glimpse of coastline and with that simple aid were able to identify the Gulf of Sirte. This gave him a compass bearing and we were lucky an hour of so later to land safely at our destination.

On another occasion I joined an RAF plane containing a group of brass-hatted senior army officers travelling from Cairo to Nairobi with a refuelling stop at Khartoum. There I received a signal to stop-over for a couple of days for some consultations. Watching the plane depart I regretted not carrying on to Nairobi and there meeting old friends. It never arrived; but a month or so later my cousin Humphrey Slade, who lives in Kenya, spent a weekend walking over the Aberdare Mountains. This range rises to some 3500 metres north of Nairobi. He saw a glint of metal near the summit and going there found the plane with its contents. It had flown straight into the side of the mountain, the pilot having presumably misjudged the timing of the descent to the airfield at Nairobi, when the Aberdares were hidden in cloud.

Apart from a few episodes of this sort our mission to the Middle East countries proceeded smoothly, with great interest, and I hope with advantage to the region. It is impossible here to give a comprehensive account but a few examples may suffice. These examples start with some impressions from Palestine, Cyprus, and Jordan, go on to international rivers such as the Nile, then turn to the many uses of water, including fisheries, and conclude with human diseases and the population explosion.

Since a large part of the Middle East is desert, and even the Mediterranean coastal lands and the mountainous areas suffer long dry seasons, problems of development in the different countries showed a good deal of similarity, the availability of water being generally a major factor. Moreover, since the bulk of the populations follow Islamic religion in its various forms there was also some similarity in the human problems. The outstanding exception to the latter was of course Palestine most of which later became Israel. The Jewish scientists and technicians who had settled in that country since the Balfour Declaration of 1917 had been greatly reinforced by the influx of refugees during the war. They had already placed Palestine in a unique category, far ahead of other countries in the region in modern scientific thought and understanding, and in all activities depending thereon. Our mission's first visit was to that country, and picking the brains of leading scientists and the British colonial officers helped greatly to set the scene. The process was backed by visits to Arab villages and to the remarkable Jewish agricultural settlements which were experimenting at that time with every variety of social organization from the individualistic approach, through various levels of co-operation, to extreme forms of sharing in work, recreation, and family life. Before going there I had experienced an hour in London with Chaim Weizmann who at the time was occupying a suite in the Dorchester. He did all the talking, about the future Israel as he saw it, and on emerging into Park Lane I found myself wondering whether I had been with Abraham.

Another early visit was to Cyprus, which, though in some ways highly atypical among the Middle East countries as far as human population is concerned, exhibited, as is the way with islands, many of the physical and biological problems of the mainland in a relatively tiny area. Bernard Keen and I arrived there four days before Christmas 1943 and the then Governor, Sir Charles Woolley, immediately called an evening meeting of heads of his departments together with H. M. Foot (Lord Caradon) who was in process of taking over the post of Colonial Secretary. The others were R. L. Cheverton (Medical Services), C. Raeburn (water and geology), R. R. Waterer (forests), J. MacDonald (agriculture), R. A.

Godwin-Austin (lands and surveys). This team of directors showed full co-operation and Foot remarkable understanding, so the upshot was that at nine o'clock next morning the whole group of directors, having handed their departments over to deputies, piled into a couple of old motor cars, with Keen and myself, for a three-day tour of the island.

In spite of wartime restrictions this group in the comparative isolation of Cyprus had addressed themselves in no small way to post-war development of the island's natural and human resources, with at the same time a proper regard for conservation of wild flora and fauna, historic monuments, and archaeology. Thus Raeburn, who had wide experience of water geology and engineering in the dry areas of Nigeria, had set about water conservation: small dams and weirs had either already been constructed or were planned for nearly every stream, however small, and through MacDonald's department of agriculture many small areas of irrigation had been developed or improved as a result. Some of Raeburn's Cypriot friends called him 'St George who was destroying the dragon of drought'. He pointed out that more murders took place in Cyprus on account of water than of women, but whether the proportion was reversed as a result of his efforts history does not relate.

Waterer, who had already wide experience of forestry in several colonies, was concentrating on conservation of that resource. He proceeded later to the directorate of Kenya and as adviser to the Colonial Office in London. In drier parts of the Middle East woody vegetation is generally cut and browsed heavily, and the forest officer tended to regard it as a moment of achievement when he could look up rather than down at his trees. The pattern of reserved and unreserved forests in Cyprus was much like that, except high up on the Troodos mountains which showed some good stands of pines and one or two surviving areas of the famous Cyprus cedar. Tree seedlings had been decimated over many years by unrestricted goats, but Waterer and his staff were able to change that by regulation and propaganda, labelling goats as 'black devils'. They were able to show some beautiful examples of a few acres each of a multitude of young cedars surrounding one or two ancient seed trees.

Meanwhile Cheverton heading up the medical department, and his medical entomologist, M. Aziz, recognized that the physical and mental capacity of a good proportion of the people in Cyprus was suppressed by endemic malaria. They were planning elimination of *Anopheles* from the entire island, a project which was carried out during the post-war years with a high degree of success and became an example for a number of similar projects elsewhere. Connected with all these and other subjects was the overriding problem of land tenure and the detailed mapping which must go with it, and this was the province of Godwin-Austin. During this tour of the island we stopped whenever opportunity offered at the numerous and unique sites of antiquity. They ranged from well-preserved Neolithic villages through Bronze Age, Iron Age, classical times, and on to Crusader, Venetian, and Turkish, all within the compass of a motor drive. We were much impressed with the opportunities for tourist development, which indeed took place after the war.

The experience of this rapid tour of Cyprus, coupled with that in Palestine, focused the mind wonderfully on what appeared to be the major problems of Middle East development. It established a pattern into which could be fitted many subsequent experiences during the mission. Our first excursion from base in Cairo included also Trans-Jordan with its almost purely Islamic culture—contrast to the mixed Christians, Muslims, and Jews of Cyprus and Palestine. In Trans-Jordan the range of natural resources and the physical and biological problems relating to them showed a similar pattern, but human pressure was less and there were relatively few specialist staff. That country in fact provided a good introduction to the larger and more dispersed Islamic countries such as Iraq and what is now Lybia.

From these initial impressions emerged some common themes which ran through the entire region and one of these was clearly water supply as related not to a single town, a rural area, or a country, but to a whole catchment area, whether or not the river which drained it finally reached the sea, or was used up by irrigated agriculture, or merely dried up in the desert. Each catchment area included all the subjects, whether human, biological, or physical with which ecology and development were concerned. Moreover most of the rivers, passing as they did through several countries, could become influences towards political and economic cooperation, or of discord.

The seven rivers of most importance were the Nile, Euphrates, Tigris, Karun, Jordan, Orontes, and Litani and all but two of these were shared between a number of different countries. Thus the Nile, of which the waters were used mainly by Egypt and the Sudan, had its headwaters in Ethiopia, Eritrea (then a separate country), Kenya, Uganda, Tanganyika, the Belgian Congo, and Ruanda-Urundi. The Euphrates, used mainly by Iraq, rose in Turkey and flowed through Syria. The Tigris involved four countries, Turkey, Persia, and Syria, before flowing for the greater part of its length in Iraq. Three countries were interested in the Orontes, namely the Lebanon, Syria, and Turkey. The Jordan was shared by Palestine, Trans-Jordan, Syria, and the Lebanon. Only the Litani in the Lebanon and the Karun in Persia were limited to one country each, and even with them international interests existed: in the case of the Litani because of suggestions that a part of its water might be deflected into the dry lands of northern Palestine, and in the case of the Karun by reason of the great quantity of silt which it poured into the Shatt-el-Arab. It was quickly

Fig. 14. Swamp Arabs of Iraq in their coracles.

apparent that each of these seven rivers, and many others of lesser importance for water supply and irrigation, had their own special problems in hydrology, control, and use, as well as some overriding considerations common to them all.

My own preference among these international rivers was the Nile, for several reasons; I knew its headwaters in the great lakes of East Africa; of all the great rivers of the world it was one of the best known from the hydrological viewpoint; and I had admiration for, and friendship with, Dr H. E. Hurst, the greatest authority on that river. At that time six volumes of his work *The Nile Basin* prepared in co-operation with P. Phillips and R. P. Black were published and I was a little proud that in the first volume I was recorded as having made the only measurement available at that time of the flow of water down the Semliki river, a Nile tributary which drains a considerable part of Mount Rouwenzori as well as Lakes Edward and George. The measurement was made by my wife and myself in 1931 when camped at the outfall of the Semliki from Lake Edward, by the simple process of floating bananas (the proverbial orange was not available) down a measured distance timed by stopwatch, after the cross-section of river had been sounded and its area estimated.

The ancient Egyptians first recorded river levels four or five millennia ago on nilometers some of which can still be seen today, for example on the Roda Island gauge at Cairo. These ancient nilometers recorded a fairly complete series of maximum and minimum levels from AD 641 to AD 1450. More recently water levels had been recorded at Aswan since 1870 and

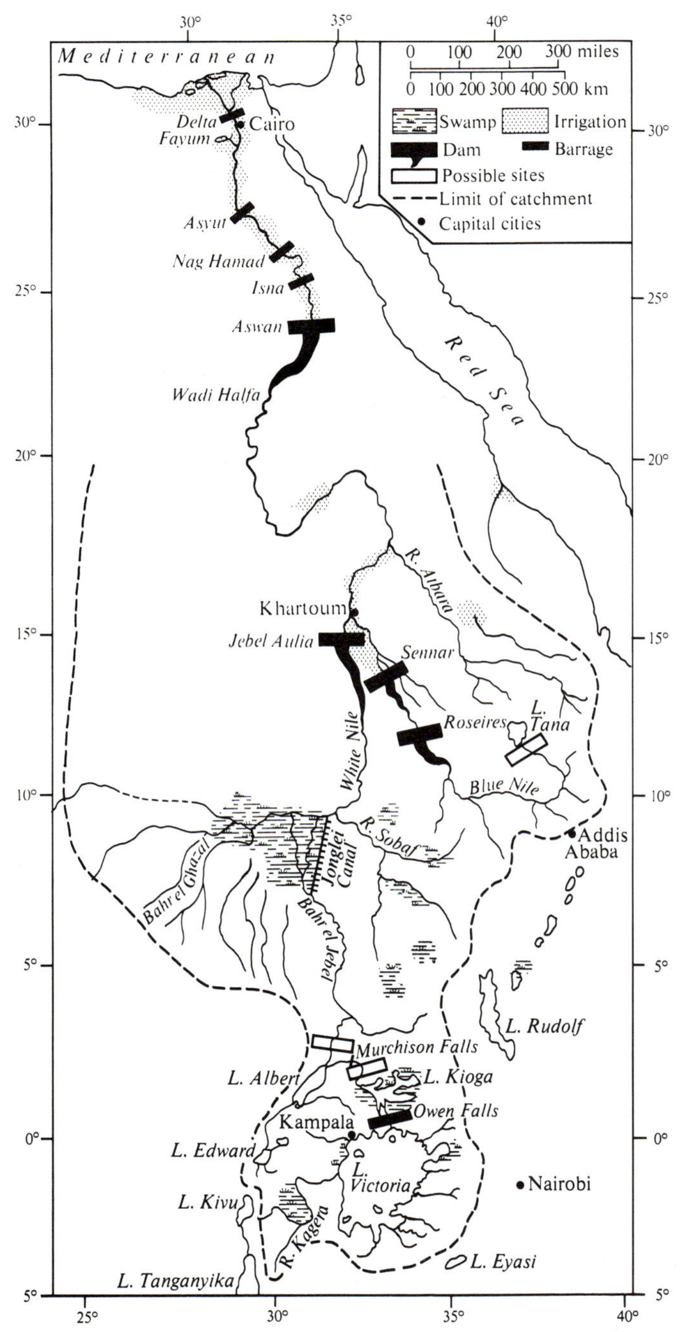

Fig. 15. Sketch map to show water control of the River Nile.

during the present century the studies upstream to the river's many sources had been greatly intensified. Quite a lot about the river's natural history—fishes, birds, reptiles, mammals, and vegetation—as well as human activities and artefacts such as boats, is also recorded in Egypt's ancient monuments; but the comprehensive ecology of the river was at the time of our mission little understood. One of those who stimulated this interest in later years was Julian Rzoska, already mentioned in Chapter 4, who established a unit for aquatic ecological study at the University of Khartoum just after the war. Another prominent ecologist much concerned is Mohamed Kassas, formerly Professor of Botany at Cairo and active in the international scientific scene.

Hurst presented me with a copy of his *Short Nile*, which compressed his multi-volume work into 100 pages or so; it was printed on the backs of old maps because there was no paper available in Egypt at the time. He then arranged a tour of the Nile in Egypt, from Aswan to the Mediterranean, visiting the barrages and other points of hydrological importance in company with the local inspectors of hydrology and irrigation. During the evenings in comfortable resthouses Hurst explained his theory of century storage, covering climatic variability. His reseach had extended, for example, to the measurement of the annual rings of big trees in California. He explained the long-term plans for further projects of Nile control, including a dam below Lake Albert, a barrage below Lake Kioga, and even control of the flow from Lake Victoria. These discussions, in which the natural history of the river figured nearly as much as hydrology, had influence a few years later when it came to proposing the Owen Falls Dam at the outfall of Lake Victoria (see Chapter 6). The Jonglei canal project, now in 1981 under construction, was also first proposed at about this time; its first detailed study was conducted during the latter days of British rule in the Sudan.

Under Hurst's ingenious influence the workshops of his Physical Department in Cairo constructed most of the precision instruments needed in the ever-continuing study of the Nile's hydrology, and it played a useful part also in the war effort. For example, shortly before the battle of El Alemain a consignment of slide-rules demanded by GHQ had finished its journey from England at the bottom of the Mediterranean. Slide-rules were essential tools of every gunner in those days before the pocket calculator, so Hurst was called urgently by GHQ: 'Could you make a lot of slide-rules immediately?' He answered 'Yes', and very soon delivered. Anyone who has used a slide-rule knows just how accurate the markings on the several scales have to be: the Egyptian physical department's version worked well and I treasure the example that Hurst gave me during our Nile tour.

Hurst handed over the department to his Egyptian colleague Dr

Simaika in 1946 after 40 years' service in Cairo, but he was then retained as hydrological consultant for another 15 years or so. With his former second in command R. P. Black he spent several months each winter in Cairo checking records of the river's behaviour.

The increasing population of Egypt compelled a continuing and ever-growing need for more water, a need which was also making itself felt strongly in the Sudan. The only obvious possibility of meeting this need was through more storage of flood water for release at the time of year when the flow of the river is at low stage. On general ecological as well as engineering principles, such storage is best provided by dams in the upper reaches.

On the White Nile system, which has a fairly constant and steady natural flow owing to the great lakes and swamps upstream, additional storage had, by this tune, already been provided by the Jebel Aulia Dam where the reservoir was filled to maximum level for the first time in 1943, but there were additional opportunities for major storage further upstream. There was also the possibility of reducing the 50 per cent loss of water from the Bahr El Jebel as it flowed through the sudd area between Juba and Malakal. The Blue Nile, which is subject to very large seasonal variations in flow and silt content owing to the heavy rainfall and soil erosion in the Ethiopian mountains, was already partly controlled by the Sennar

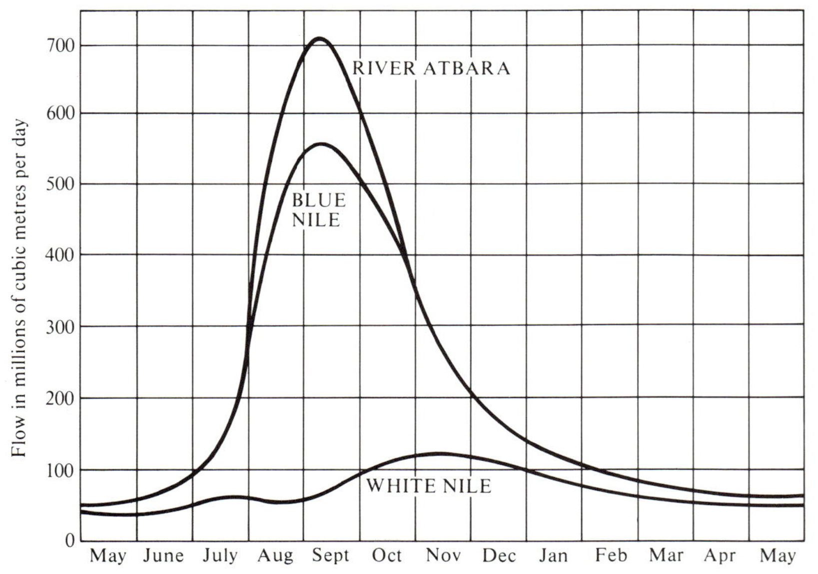

Fig. 16. Average flow of the River Nile at Aswan to show contributions to the total from the Blue Nile, White Nile, and the Atbara.

Dam, completed in 1925. The prime purpose of this was for irrigation in the Gezirah area of the Sudan rather than to increase the supply to Egypt. The possibility of a dam at the outlet of the Blue Nile from Lake Tsana had been the subject of several missions but did not look promising for political reasons. There was, however, the possibility of a major dam and water storage at Roseires a little downstream of where the Blue Nile enters the Sudan.

These various projects made up a pattern of river control of which the hydrological needs and consequences had been worked out conforming to Hurst's principle of long-term storage designed to provide for cyclical or secular changes in climate as well as the pronounced annual variation of Nile flow.

During later years, with the exception of the Owen Falls dam which now controls the outfall from Lake Victoria, and the Roseires dam in the Sudan which provides storage for a part of the Blue Nile annual floods, these plans were put on one side when the Russians offered to build the High Dam at Aswan. This great project had certain obvious disadvantages, for example by retaining the fertile silt from the Blue Nile floods which, from the old Aswan reservoir could be flushed out each year and passed down stream to the irrigated fields. Nevertheless the amount of extra water stored in Lake Nasser above the High Dam has enabled a very large area of Egyptian desert to be brought under irrigation and it has practically reduced to nil the loss of Nile water to the Mediterranean sea. Hurst's opinion was initially against the High Dam, but by 1970 when I called on him in Cairo with a mission from the United Nations Development Programme advising on environmental research to be undertaken in connection with the High Dam, he had come round to agreeing that it was right. But he predicted that before many years had passed the upper Nile projects would once again come to the fore for serious consideration. This indeed has already happened. The Jonglei canal, for example, will conduct about half the flow of the Bahr-el-Jebel in a canal to the east of the river's natural channel with a minimum loss from transpiration. When completed it should allow about three-quarters of the flow above the sudd area to become available for use downstream, instead of the present one half.

The Nile has of course always provided a route of contact between the Mediterranean region and tropical Africa, not only of human beings but also of plants and animals which had been evolved respectively in the Palaearctic biogeographical region to the north and the Ethiopian region to the south. The Sahara had been one of the world's main barriers to the mixing of genes. Meantime for some 5 millennia Egyptians had been the dominating human influence on the Nile and were wholly dependent on its waters. At the time of which I write the use of its water was controlled by the Nile Waters Agreement of 1929 which acknowledged fully 'Egypt's natural and historic right to the waters of the Nile and its requirements for agricultural extensions'. As the Governor of Uganda once remarked to me a little irritably when, as part of that country's development plan, I suggested constructing a dam at the outflow from Lake Victoria: 'But what will Egypt say? They can use the Nile's water just as they like, but here we can scarcely drink a cup of tea without their permission.'

Other international river catchment areas have stories comparable to that of the Nile, though less complete in the historical sense, but our mission paid attention also to the small rivers, some of which, now reduced to wadis of intermittent flow, had been of importance to past civilizations and might perhaps be used again. Along the coast of Tripolitania a number of wadis which rise in the high Jebel Nefuza and flow northwards to the Mediterranean, carry large quantities of water and silt each winter but are dry for more than six months each summer. It is near the mouths of

Fig. 17. Water lift from River Orontes: a *Noria* under repair at Hama in Syria. The river turns the great wheel which lifts water to the high-level aqueduct seen in the distance.

such wadis that the three great cities of Roman times, Leptis Magna, Oea (Tripoli), and Sabratha (City of the Fish) were situated. Travelling up the wadis behind these ancient towns, which had been partly reconstructed by Italian archaeologists before the war, it was at once apparent how admirably the Roman system of masonry dams was adapted, not only to provide permanent reservoirs for the towns themselves, but to counteract soil erosion far up in the tributaries.

For example, the tributaries of the Wadi Cam, which, when it flows today, reaches the sea ten miles east of Leptis Magna, show the ruins of dams constructed of Roman masonry at intervals of less than a mile. No doubt the reservoirs contained by them were silted up fairly rapidly, but they served as settling tanks to reduce sediment in the larger reservoirs of the lower reaches, the last of which took water by an aqueduct, partly underground, to Leptis Magna itself. It seems that, in addition to the silt-catching dams upstream, several reservoirs were constructed in the lower reaches of this wadi, and during the long dry season their water could be

Fig. 18. Dam across the Wadi Hadhramaut being constructed by traditional methods. It is designed to raise the level of the underground aquifer as well as to divert flood water for irrigation.

released one at a time to keep the lowest in the series constantly filled. During the pre-war Italian colonization one or two of the ancient dams near the wadi's mouth had been reconstructed, but the reservoirs so formed were of little value because silt accumulation was extremely rapid. The Italians had neglected the Roman principle of bringing the rivers under control first at their headwaters, a mistake which has been repeated many times in the vast expansion of dam building all over the world during recent decades.

Wherever water is available it is used in many ways: its uses for domestic supply, for domesticated animals, for irrigated agriculture, for fisheries, transport, and the disposal of waste are obvious enough. Less obvious, until the science of aquatic biology became well established about the turn of the century, was the fact that many organisms which cause human and animal diseases use water as a transport system from one host to another, either direct or via aquatic animals like mosquitoes and snails —as vectors. The proximity of water was therefore not always an advantage from the human viewpoint.

For the immediate purposes of MESC in reducing imports for civilian purposes the use of water in food supply, that is in agriculture and fisheries, was of paramount importance. The many different agricultural systems adopted in the Middle East, studied by Bernard Keen during our mission, revealed various ways in which improvements could be introduced with advantage; but they also showed how well many seemingly primitive methods were adapted to local ecological conditions. To take an example, many people had drawn attention to the apparent waste in using the dung of domestic animals as fuel for cooking; but when dung is spread on dry land at a high temperature its oxygenation is so rapid under a powerful sun that its fertilizing capacity is largely lost. Thus, although beneficial as manure at times of rain and under irrigation, in dry farming it may be better to draw heat from dung in the kitchen, and thereby to reduce the pressure on wild shrubby vegetation which in many places is the only alternative fuel. Another example is the simple nail plough still used in Arab countries, which scratches the topsoil sufficiently for seed to take root, but does not disturb the subsoil. Its replacement by the deep-delving European plough has sometimes caused disastrous soil erosion.

The study of fisheries, which had been somewhat neglected previously in this region, ran in parallel with that of agriculture. Most fisheries, being based on natural regeneration, require the minimum of attention in order to maintain their yield, provided over-exploitation is avoided. Fishermen are thus comparable more to hunters than to cultivators in their mode of life. Even when they over-use nature's bounty an area of water usually recovers more quickly than does soil. At the time of our survey there was however one branch of the fishing industry, then quite small but now much expanded, namely farming fish in ponds, which involves processes comparable to tilling and fertilizing the land, sowing, weeding, and cropping.

Of the seas around the Middle East countries, the Mediterranean did not offer much opportunity for expanding fisheries because, as sea water goes, the Eastern Mediterranean has a very low productivity. Evaporation is so high, and the amount of fresh water received from surrounding countries is so small, that the Eastern Mediterranean's water is largely replenished from the west, and during the slow flow eastwards from Gibraltar most of its nutrient salts are consumed. From the fisheries viewpoint the main redeeming feature was the river Nile which, in a state of nature, used to pour rich silt-laden waters into the sea, building the great delta in the process. This resulted in high, though localized, production of sardines and a once flourishing fishery which reduced as the progressive stages of Nile control allowed less water to reach the sea. The Eastern Mediterranean was well known however for its sponge fishery which extends from the coast of Lybia right round to Turkey. The tunny fishery by traps off the north African coast, which was limited to a short season of six weeks in June and July, had already been over-exploited. The deep coral-fringed waters of the Red Sea produce a good deal of fish for local consumption, but they too suffer from impoverishment of nutritive salts and hence no major commercial fisheries.

To witness really important marine fisheries, underexploited and with great promise for the future, one had to travel out of the Red Sea into the Gulf of Aden, where upwelling currents carried rich deep water of the Indian Ocean near to the surface. This fishery of the south coast of Arabia is seasonal, dependent on north-east winds in winter and the south-west monsoon in the summer. During the winter sardines come inshore and are caught in vast numbers by castnets. This indeed is one of the few parts of the world where in normal times fish is in superabundance. The coastal Arabs are essentially a marine people, fishing and handling the dhow traffic of the Indian Ocean, and for much of the year they appear to eat little else but fish. Even their transport camels, plying between the coast and the unique civilization of the Wadi Hadhramaut a hundred miles inland, are foddered on dried sardines. Further east, beyond Saihut, where the true 'ichthyophagi' live, even cattle eat fish. Colin Bertram, who in the latter part of our mission was appointed fisheries officer to MESC, included this area in his study and emphasized that the fishery could in future produce a large surplus of animal protein, even beyond its normal export to India and East Africa. In subsequent years a good deal was done to this end, first by the British so long as Aden was a Protectorate and later by the FAO.

In efforts to increase the fish produced we devoted particular attention to the inland waters because production from them could reach a larger proportion of the people actually living within the Middle East, many of them suffering from protein deficiency. Reservoirs on the Nile and on the Tigris and Euphrates were capable of producing good quantities of fish, and so were the natural lakes and swamps of the sudd area of the Nile and the confluence of the Tigris and Euphrates, the former exploited by Shilluk and Nuer tribes the latter by Swamp Arabs. The delta lakes of Egypt and Lake Karun also had fisheries of importance; these had been enhanced by the import of elvers each year from countries further to the west with the resultant eel fishery. It was however the fish farming industry, mainly for carp and *Tilapia*, which was at the time being pioneered by the Jews in Palestine, that offered the greatest opportunity for expansion.

Swamps, fish ponds, and irrigated agriculture lead on to considerations of human health. The pattern of government services in each country indicated a concentration on obvious short-term objectives rather than on those needs which might be more beneficial to social and economic progress in the long term. Thus disease in human beings is more obvious than damage to the environment caused by deforestation, inappropriate systems of agriculture, and soil erosion. For this reason in underdeveloped countries it is usual for medical services to be first in the field, to attract more money, and to advance more rapidly than services designed to improve production. The appeal for curing diseases had sometimes left their underlying causes intact. Curative medicine generally had progressed more than preventive medicine. With a few outstanding exceptions such as the Jewish agricultural settlements in Palestine, the third stage, that of building up a population of healthy people through an overall rise in nutrition and other living standards, had hardly yet progressed beyond the planning stage.

Of prevalent diseases malaria was by far the most important in the Middle East at the time of our mission. Malaria was the fundamental reason for the backwardness of many village areas and without a much higher degree of control, progress in education, agriculture, and social welfare was slow owing to the general debilitation which it caused. In arid areas, where malaria had not been a problem owing to lack of water for mosquitoes, serious epidemics of malaria regularly followed the installation of irrigation. Fresh in people's minds was a very serious epidemic in upper Egypt when *Anopheles gambiae* had invaded from the Sudan. Soon after the war, DDT came as a godsend in the fight aganst malaria, but its heyday lasted for only a couple of decades until the environmental revolution got into its swing (see Chapter 11).

The other main debilitating disease of wetland areas was schistosomiasis (bilharziasis) of which there are two kinds in the Middle East and Africa, urinary and rectal. Both are transmitted by certain kinds of water snail, so the problem of control is different from and more difficult to solve than that of malaria. It seems that schistosomiasis has expanded its range and incidence rapidly during the past half century. Around Lake Victoria this disease was practically unknown to medical staff at the time of our fishery survey in 1927, but soon after the war it was rife in nearly every fishing village. This, it would seem, was brought about by a great increase in the number of Africans travelling from one place to another, from the irrigated areas of the Sudan, for example, where the incidence of schistosomiasis approached 100 per cent of the population, to Uganda where it was minimal.

Schistosomiasis is caused by parasitic worms living in the human blood-vessels and the worms' eggs reach the external environment via urine and faeces. In water the eggs hatch into larvae which penetrate into and develop in the snail. Later a second type of larva hatches from the snail and enters human beings by boring through the skin. Ultimate control will undoubtedly come by preventing human urine and faeces being deposited in or near water which contains the snails, but that demands a high level of sanitary and social development so that a variety of other methods are practised. Some success is achieved with molluscicides and in irrigation schemes a good deal can be achieved by engineering design in avoiding contact between the human being, especially children, and water which may become infected. At the time of our Middle East study schistosomiasis was recognized as widespread but was apt to be taken for granted. After all it had been diagnosed in Egyptian mummies and had probably been endemic ever since. But, as the full extent of the growing problem became recognized, it now heads the list of problem diseases in many warm countries. Schistosomiasis rarely kills, except in association with other conditions, but some authorities estimate that it can reduce the capacity for work of infected populations to the extent of 50 per cent.

In the overall pattern of human ecology in the Middle East at that time diseases certainly occupied a prominent place. Some of them were diseases of the environment, especially malaria, but including also in some countries yellow fever, sandfly diseases, sleeping sickness, and onchocerciasis. Diseases of insanitary conditions included the worm diseases of schistosomiasis, ankylostomiasis, and tapeworms, and louse-borne diseases of typhus and relapsing fever; there was also cholera, typhoid dysentery, TB, leprosy, and the very common trachoma of the eyes. Smallpox was still prevalent. Veneral diseases which are among the few which the individual had the power, if not the will, to prevent, were also very prevalent. Some doctors with long experience of these countries considered that some of the most regrettable conditions were caused by female circumcision which in various forms was widely practised by most Islamic and some other communities. Recent accounts appear to indicate that there has not been much reduction in these practices during the past 35 years.

Another subject constantly in mind was the increase of human populations. Among the developing countries of the world Egypt is one of the few which has firm historical records, and the story of its increase in population is a classic. About 1800, in Napoleon's time, the country was estimated to contain about 2½ million people. Subsequent censuses arranged by the British showed that the number had doubled steadily each half century, so that by the end of the Second World War it was pressing on 20 million. At the time of our survey this was causing serious alarm, but since then the number has about doubled once again. The current population explosion was probably not, however, the first that Egypt had suffered, for ancient history shows cycles of prosperity followed by disaster which may well have been associated with large fluctuations in population. On this subject that great scholar W. L. Balls who, more than anyone was responsible for the early prosperity of the Egyptian long-staple cotton industry, made a striking contribution to a Middle East agricultural conference which MESC convened in Cairo in 1944. He suggested that there had been a population explosion among negro slaves shortly before the building of the great pyramids; that the surplus of slaves was put to the massive task of pyramid-building but labour did not unfortunately reduce their birth rate. There was revolution and during the subsequent period of some 400 years the historical record of Egypt is almost blank. Although some Egyptologists may not agree with that interpretation Balls made his point in warning for the future.

Looking back at the short but exciting period of this scientific survey one could reach certain generalizations as well as making specific suggestions. Of the four stages in human development—hunting, pastoralism, cultivation, and urbanized industry, hunting continued only as a recreation except in a few areas, mainly on the African continent. Pastoralism occupied by far the greatest area of land and extended far into the deserts. In principle it was ecologically sound and often the best way of using limited and localized resources of biological production and of water, but in many countries the pastoral population had reached a stage of unbalance as pressure on the land had increased. In agriculture of the more arid areas, dry farming by traditional methods relying on rainfall which might never come, was also a remarkable adaptation to local conditions. To meet the risks of failure of the rains it had to be combined with storage of grain. Irrigated agriculture, which in several countries dominated production of both local food supplies and exports, offered a great opportunity for increase, but it was fraught with a multitude of technical problems which in past ages, and even at the present time, had resulted in about as much land being vacated through salinization of the soil and diseases as was brought under fresh cultivation. This was where science—physical, chemical, biological, medical, and social—had the greatest part to play. The truly industrial stage of development was at that time only beginning, but it was widely recognized that its rapid development would have to provide the short-term solution to the human problem in some countries which were already over-populated.

The problems resolved themselves into those aiming to remove factors which limit the extension of existing means of livelihood, and those aiming to change the means of livelihood in order to counter population increase by something other than disease, premature death, and general inefficiency. An improvement in the standard of living, education, and health services, though it might untimately lead to a stable population, would have a long intermediate stage when it tended to increase rather than to reduce the rate of multiplication.

Since progress in one direction was bound to have repercussions in many other directions, the survey was arranged to show the interrelation of subjects, and from this two special foci emerged—water supply and population dynamics. Considering the current economic and political concentration on oil, that may sound surprising, but at that time oil was looking after itself well although there were complaints from those concerned with other underground resources that the extensive geophysical and geological knowledge derived from oil prospecting was kept so secret.

Apart from oil, planning in all subjects had been held back by lack of financial resources, political uncertainty, and lack of fundamental knowledge about the problems. In most countries the disparity between rich and poor was glaring as a social problem and a brake to planning for a better quality of life. Subsequent reforms by independent governments, and great wealth from oil, do not seem to have solved that problem yet. There was particular need for experiments in planning for rural areas before large-scale operations were undertaken with grant-aid, and this necessitated pilot schemes which concentrated facilities and money in small sample areas, even though this ran counter to he usual principle of even distribution of advantages. Although there was obvious scope for planning on a large scale, for example on the scale of a whole river's catchment area, development was generally more successful coming from below, from the grass-roots, than imposed from above. Thus importance attached to planning with the village or even with the individual home as the unit.

The greatest need of the Middle East appeared to be a deeper sense of public responsibility. During the war a sense of cohesion in scientific and technical subjects had been fostered, and we who had become involved hoped, at the time, that the organization would continue in some form as an Anglo-American effort for a few years in order to lay a foundation for cooperation in post-war progress. To this end Dunstan Skilbeck, who was at the time head of the agricultural group in MESC and later became Principal of Wye College in Kent, was particularly active. Planning was handicapped by political boundaries, but in 1944 Stilbeck organized an agricultural conference with participants from 17 countries: a joint programme of research was prepared to be based on a Middle Eastern Research Centre. He even achieved the allotment of funds from several of the countries, but in the event these and other plans became submerged in political turmoil as the war receded. Nevertheless the work of the scientific mission was not wasted. After America pulled out the British established a new headquarters for a Middle East Office at Beirut in the Lebanon. Paul Howell, an administrator from the Sudan service who had been in charge of the Jonglei ecological investigation, was in charge of scientific and technical assistance, and he was able in subsequent years to put into operation some of the recommendations we had made. Another post-war development was the establishment by Unesco of a scientific office in Cairo with Michel Batisse, as its first director. More than 30 years later, when Batisse, then head of the Natural Resources Division of Unesco in Paris, was helping to make the International Biological Programme work, he brought me his copy of *Middle East Science* for autograph, saying how it had provided initial guidelines for the wide-ranging activities of Unesco in the region.

As a postscript to this chapter, I may add that in October 1977 the United Nations Economic Commission for Western Asia convened a conference at Beirut on 'Technology transfer and change in the Arab world' (Zahlan 1978). As one of the small group of non-Arabs invited to attend, I explained the situation as it had appeared to our mission in 1945 and attempted to trace some of the changes which had taken place since then. The really outstanding change became evident at the conference itself. In 1945 it was difficult to find any indigenous scientists or technicians with whom to discuss the many problems of technical transfer and change. The 1977 conference bubbled over with knowledge of such problems and with ideas and plans for solving them.

6
Uganda (1946)

Uganda . . . is alive by itself. It is vital: and in my view, in spite of its insects and its diseases, it ought in the course of time to become the most prosperous of all our East and Central African possessions . . . My counsel plainly is: Concentrate upon Uganda . . . Nowhere else will the results be more brilliant, more substantial, or more rapidly realised.

Winston Churchill (1908) wrote that when he was Secretary of State for the Colonies and had been on a tour in Africa. It was a prophetic statement and was well proved by history up to the takeover by Amin in 1971.

This small country in the middle of Africa was bound to have a prosperous future, politics apart, owing to the richness of its natural resources. Resources of climate, soil, and water provide great protential for agriculture and fisheries, timber, wild life, and it has mineral resources as well. The people, too, after a turbulent past, had become relatively diligent. The country had been isolated some 400 miles from the sea in a beeline, but by the time Churchill wrote there was an eastern outlet by lake steamer across Lake Victoria to Kisumu and thence by rail through Kenya to Mombasa. There was also a longer route southward across the lake to Mwanza and thence by railway to Dar-es-Salaam, and another route down the Nile to Egypt with nine changes between boat, rail, and road.

Considering its comparatively small area, Uganda is more varied geographically than most African countries. A fifth of its area is lake or swamp and it contains the greater part of two of the highest mountains in Africa, Ruwenzori to the East and Elgon to the West. Such features account for an agreeable climate with ample rainfall, although in the north-east it verges on aridity. With such variety of habitat the country's richness in wild plant and animal life is almost unequalled. The human inhabitants, too, are diverse, derived from Bantu, Nilotic, and Hamitic roots, and including well advanced and very primitive tribes. At one end of the scale were the Baganda who had evolved a highly complex social system dominated by the Kabaka. At the other end were small isolated peoples living almost in the stone age, such as a group of fishermen called Bakenyi who lived on the papyrus swamps of Lake Kioga and sometimes were moved by water and wind on floating islands, complete with their huts and canoes.

Churchill's visit was only a few years after Sir Ronald Ross had shown malaria to be transmitted by mosquitoes and Sir David Bruce had announced that sleeping sickness is carried by tsetse flies. Uganda was not a healthy country then. Some 200 000 people had died from sleeping sickness along the northern shore of Lake Victoria, while malaria was endemic everywhere. No memorial stands for the African people, but there are many sad little cemeteries recording the death of British officers and missionaries, many of whom were cut off pathetically soon. However, discovery of the causes and of the vectors of diseases had opened up means for their control, albeit not complete even to this day.

It was to this fascinating country that I was summoned in March 1946 by its new Governor, Sir John Hathorne Hall, in order to prepare its first ten-year development plan which was required under the Colonial Development and Welfare Act of the British Parliament. Such a plan, even if accepted and carried out in full, would in the end become but a brief episode in the history of Uganda, but it would set a precedent and so presented a dilemma to an ecologist like myself. It would be rather easy to put together a plan on purely economic grounds—aiming at rapid development, even exploitation, of natural resources. It would be equally easy to focus on the people and concentrate on improvement to health, education, and social progress. But what of the long-term future? There was, for the moment, some extra money available from Colonial Development and Welfare (CD&W) but that was intended to initiate new developments, not to pay for them indefinitely. Before long Uganda, like other colonies, would have to become viable on its own, and that implied paying for increased services to the people by increased production and trade. The known resources for increased production were almost all biological and as such were self-regenerating in theory; but I knew enough about tropical ecosystems to realize that in practice many of them were brittle and easily damaged. There was often temptation to exploit rather than to develop a resource; to live on capital rather than income. A tropical forest, if cut down and sold, would take centuries to regenerate if it was ever allowed to; some of Uganda's soils were among the richest in Africa, if not in the world, but it was improbable that their richness would survive continuous cropping. Clearly conservation must become an integral part of development. There were dangers to the people as well as to natural resources. At that time some 90 per cent of Uganda's population, like in most of Africa, lived in rural environments in close contact with nature. The processes of urbanization were only at their beginning. But I had seen enough of big cities to realize how the quality of life can be eroded by them, and to be somewhat frightened at the prospects of population increase leading to industrialization as advocated by most economists at that time. Thus while the overall objective of the plan must be the future well-being of the people and their country, a note of caution

Fig. 19. The black baby.
Mr Bull: 'What another!!—Well, I suppose I must take it in!!!' (From *Punch* 21 April 1894.)

was needed as well as a maximum integration of the many subjects involved.

I first looked at the past in order to put this dilemma into some kind of perspective. For many millennia in the middle of Africa the human species, with a tremendous variety of other animals, had been reacting with the equally diverse plant life, so that many different ecological systems had evolved. This started a very long time ago, as was being demonstrated by E.J. Wayland, the retired director of geological surveys, who was at that time still working in the field on palaeolithic stone tools. I spent a night at his archaeological camp in a remote part of the country and we talked late about the changes in the level of Lake Victoria and of the direction of river flows which his research indicated, and how they related to the distribution of fishes and the early hominid remains which Louis Leakey was then digging in Kenya and Tanganyika Territory.

Since there was no true history, that is written record, for this part of Africa until the explorers and missionaries of the nineteenth century, most of the events after those very early stone age times had to be reconstructed with some guesswork. However, it seemed very probable that *Home sapiens*, having originated in Africa, emigrated north-eastward and produced the great early civilizations in Egypt, Mesopotamia, India, and China. From time to time there were reinvasions, mainly from that direction, and also perhaps from the Far East by sea to the East African coast. Some of these waves of people, especially those of hamitic type, brought with them cattle, sheep, goats, and horses which they had domesticated in Asia; others brought technologies, religions, and forms of social organization which were alien to those Africans who had continued to live in their land of origin.

One had to suppose that the original Africans had lived in close harmony with nature, hunting and food gathering, probably with some cultivation but with no domestic animals. Perhaps the biggest effect on the land was the use of fire to reduce dense vegetation and enable people to get around. But with the Hamitic invasions and the introduction of domesticated herbivores changes doubtless became far more intense: woodlands were cleared, pastures encouraged, and beasts of burden gave greater mobility. There was soon doubtless some accelerated soil erosion near settlements, and it seems that tsetse fly became a factor of importance at that time. Some time after the seventh century AD the religion of Islam reached parts of Uganda, but the country's isolation was such that slave raiding and trading did not affect its people much until the Arab safaris penetrated the interior from their bases at Zanzibar and Bagamoyo.

Thus gradually Uganda and its people took on the shape found by the British explorers seeking the sources of the Nile, soon to be followed by Christian missionaries. By then the people in the southern half of the country were mainly of Bantu stock but of many different tribes not all friendly to each other; those of the north were mostly nilotic or hamitic. Some specialists thought that the Bantu tribes of Uganda represented a basic stock who remained in occupation after giving rise to those elements of the huge and varied Bantu populations which invaded southward, and came to dominate most of the southern half of the continent.

An important element of the people in Uganda and adjacent parts of Ruanda was provided by the tall handsome and intelligent Hamites, called Bahima, who, with their fine long-horned cattle, in some areas became overlords of the more numerous but smaller Bantu. This happened especially in the grassland area between Lake Victoria and Lake Edward. There were also, in several parts of the country, tribes classified as half-Hamites; but for progressive ideas none could compete with the Baganda people who occupied the central area around the joint capital towns of Entebbe for administration and Kampala for commerce.

From about 1880, when the country first became of interest to the British public, Uganda's historical record was complete. The more accessible parts near Lake Victoria's northern coast then became the scene of conflicts between the partisans of the Anglican, Roman Catholic, and Muslim faiths. Following the death of Kabaka Mutesa in 1884, his successor, Mwanga, indulged in all sorts of excesses. Bishop Hannington and many other followers of Christianity were murdered and the Imperial British East African Company had found it impossible to maintain their position after penetrating Uganda in pursuance of their charter, so they retired from any further operations in the country.

Then at the end of 1890 Captain Frederick Lugard arrived, having been deputed to make arrangements for future relations between the Baganda and the Company. During the next few years he had insurrections, mutinies, atrocities, and battles to contend with, but in spite of all, Uganda was the first of the African dependencies to adopt the policy of 'Indirect Rule' into which the name of Lugard is indelibly written—that is rule by the local indigenous chiefs and councils with guidance provided by British administrators.

Peace was restored, but Uganda's future remained dark, because sixteen out of seventeen of the British Cabinet voted for withdrawal from the country. In 1892 Lugard hurried to England to avert the threatened substitution of German for British rule, and the subsequent pressure on Mr Gladstone's government, especially from the Missionary Societies and the anti-slavery lobby, was such that a protectorate was finally declared in 1894. The occasion was celebrated by *Punch* with a cartoon depicting John Bull opening his door to find a black baby on the step, with a Uganda label.

In the event John Bull, aided by several generations of devoted Governors and Civil Servants, did not do so badly. The Uganda which I was looking at half a century later was peaceful, except for a little tribal raiding in the north-east, and was on friendly terms with neighbours—the Sudan to the north, Kenya to the east, Tanganyika Territory to the south, Ruanda-Urundi and the Belgian Congo to the west. Administration was widely based on a hierarchy of Provinces, Districts, Sazas (counties), and Gombololas, the provincial and district commissioners being all British, the Saza and Gombolola chiefs all African.

The economy depended for the most part on agriculture, which was nearly all of a peasant kind; family or communal holdings of land were growing cotton as the main export crop with some coffee and plenty of food crops. There were flourishing plantations also of sugar organized by immigrant Indians near the north shore of Lake Victoria, of tea in the highlands towards Mount Ruwenzori, as well as of cotton. Extensive forest reserves and game reserves had been established in areas which were not yet under human pressure. Food was abundant and remarkably cheap—bananas and eggs, for example about one cent each, with 100 cents to the shilling. There was more animal protein available than in most African countries owing to fish from the lakes and a useful cattle industry in the drier areas where tsetse were absent.

Government revenue had just exceeded £3 million in 1945 and was growing steadily, keeping ahead of expenditure. It was derived mainly from export tax on cotton, but there was a significant export also of coffee and of hardwoods. Apart from cotton ginneries and a sugar factory, nearly all in the hands of Indians, and a few sawmills and tea factories, 'industry' was confined to the cottage type. Health services and education, though not reaching everyone, were well found. Kampala was linked via Nairobi to the outside world by 800 miles of railway to Mombasa, and the road system had the reputation of being smooth-going and passable for at least most of the year.

The British Parliament had chosen July 1940 to pass the first Colonial Development and Welfare Act which allocated five million pounds a year for the purpose of promoting improvements in the standard of social and economic life of the colonial peoples. During those dark days we were under the threat of invasion and obliteration of our cities by German bombs. Nearly everyone expressed confidence that we should win in the end, but few had any idea how that could come about. Action on the CD&W Act had to be deferred, and by 1945 it was realized that the amount of funds allocated in 1940 was not enough. Supplementary legislation increased them, and the Colonial Secretary in London instructed every territory to prepare a plan designed for the best use of its share. Uganda had in fact started to make such a plan, but when H. E. Sir John Hall looked at the draft he thought it much too departmental, each director vying for his slice of the cake with little reference to the others or to an integrated programme. That was why I was there: to take a new look at the future of the country from outside and to prepare an overall plan of developments which would include advice on how best to spend the Uganda CD&W allocation.

I was to be aided by an administrative officer, Frank Latin, who, years before on one of my lake expeditions, had unexpectedly walked with a string of porters into my camp in a remote corner of the country. He later became a key person in the Uganda Electricity Board and after that represented the country in London. Together we performed a series of whirlwind tours, visiting every district, and had many meetings picking the brains of departmental officers and others. When back in headquarters at Entebbe I was a guest at Government House, discussed the problems with Sir John Hall, and met many interesting visitors. For example I recall an occasion when the Aga Khan, who had just been weighed against diamonds, came to lunch. He was accompanied by the Begum, who had been 'Miss France', and I was so overcome with her beauty that my halting French quite forsook me.

But, between such interludes we worked hard. Fundamentally the problem of development was one of human ecology, the diverse people reacting with their varied environments. Each aspect of the environment was looked after by a department of government, with the Legislative Council and Secretariat at the centre, so the plan had to cover problems of co-ordination and integration. There had to be considered also many non-government organizations such as the missions which contributed greatly to health and education, a good many privately owned plantations, and the Uganda Company which had originated from the anti-slavery movement but had become important in the country's agriculture and commerce. The Indian community had a prominent place, too, for in addition to urban development, its members had brought material goods within reach of the smallest village community by opening dukas (shops). The Indians had done much to introduce a money economy, but many of the needs and activities of the people could not readily be qualified or assessed in financial terms, because the quality of life was as important, often more important, than its quantity.

The main subjects which had to be considered in the plan formed a pattern of relationship similar to that which I had adopted for *Science in Africa* (Chapter 3). At the inorganic level the shape of the land, the rocks below the surface, and the climate was each looked after by a separate department of government; at the level of vegetation the two practical applications of plant science in forestry and arable agriculture each had a separate department likewise. At the animal level, the wild fauna was supervised by the Game Department which was responsible also for fisheries, while domestic animals had their separate Veterinary Department. Human beings were looked after by the departments of Administration, Education, Health, Law and Order, and Public Works. Insects, which had abundant opportunity of having attention drawn to them, pervaded the whole complex: they were food for other animals, for fish, and sometimes for mankind; they were pests of forestry and agriculture, and vectors of diseases.

To illustrate, the relationships one could start with a simple diagram such as Fig. 8, and then make others more elaborate, perhaps one for each of these main subjects, and sometimes one could quantify the various factors. This kind of process now supports the high-flying title of 'systems analysis' but to the ecologist of 30 years ago it was merely common sense. In whatever way we arranged the pieces our chief object obviously was to raise the standard of living of the people. It was equally obvious that the average standard of living could not rise, and public services could not expand, unless production increased at a rate greater than population.

From such preliminary thoughts stemmed an analysis of the main factors which limit development and welfare. They boiled down to (i) lack of fundamental information about the country and its people; (ii) systems of production that were incompatible with the full use of natural resources; (iii) lack of capacity or inclination of many of the people for physical and mental work; (iv) power based on the most inefficient of fuels, namely firewood. My plan aimed to solve the fundamental problem of balance between population and production by reducing these limiting factors.

In working it out some interesting points emerged. For example the best estimates we could obtain for population in 1946 was based mainly on hut tax figures multiplied by an estimated number of people per family, which varied from one district and tribe to another. The total added to about 4 million for the whole country. Some two years later the first reasonably accurate census, organized by the East Africa High Commission, gave a total of 5 million; the census men had discovered a million people whom nobody thought existed! It was essential also to estimate, or rather to guess, the rate of population increase, so as to have a figure to plan for at the end of the decade; and here again subsequent study showed that the figure we chose was too low.

Water supply in rural areas was constantly a difficulty, even in Uganda with its liberal rainfall, for in the dry season many women had to walk five miles or more in order to collect and carry back their three or four gallons in a calabash or petrol tin. We set the target of a permanent water point within two miles of every village, and this was just about achieved by the end of the decade through excellent work by the geological department sinking boreholes and wells, and constructing dams. Even today there is yet far to go in this highly important public service, so that the catchphrase of 'potable water for all people by 1990' was to the fore at the United Nations World Habitat Conference of 1977 and was reinforced by the Water Conference of 1978. Food production through the expansion of agriculture and fisheries, the latter having special opportunities in so well watered a country, also came in for much attention.

The social services, especially health and education, had fared rather better than the productive services in the past, and some adjustment seemed needed, but social services had clearly to expand more or less in parallel. Some emphasis was laid in the plan on education of girls for whom, as in other countries at that time, schooling provision was meagre. At the apex of the educational pyramid Makerere College was the main focus and soon afterwards became the first University in East Africa. Attention was also directed to the little Uganda Museum at Kampala, a bud of culture. Under its director who was a musicologist, it already had a fine collection of indigenous musical instruments which were played to visitors by trained African staff.

The services of administration, law and order, a statistical office and communications came into the plan also. In the transport programme it was estimated that 60 miles was about the maximum distance that produce could be moved economically between point of production and a railway station or port on one of the lakes; so certain branch railway lines were proposed, including one from Kampala far to the west to the foothills of the Ruwenzori Mountains. The primary reason for this was for the export of minerals, not agricultural produce, for a large deposit of copper ore had been located near a remote village called Kilembe to which the only communication was a very long and winding road, quite incapable of carrying this heavy rock to railhead at Kampala. Until then Uganda had but little mineral industry, other than gold panned in a few rivers and soon worked out. The branch line to Kilembe from Kampala was built in due course, with substantial benefit to the country as a whole.

Another part of the plan to which I, as a biologist, attached weight was the conservation of wildlife. This had to be approached rather delicately because at that time many people regarded wild animals as enemies rather than assets, and the number of elephants which were killed each year to protect agriculture from their depredations ran into thousands. However, Uganda was lucky in having at that time an outstanding chief game warden, Captain Charles Pitman, who was already responsible for a number of game reserves and had done much to popularize African wildlife in a series of books. He and I worked out proposals for establishing two National Parks, the idea of which was then embryonic in tropical Africa, and these were duly incorporated in the Plan. One was to be around the Murchison Falls on the Victoria Nile, an outstanding area of great scenic beauty and abundant fauna; the other was in the area between Lake Edward and Lake George which is adjacent to superb scenic country pockmarked with extinct volcanic craters. This was well before the fashion for African safaris had got into its stride; there were no minibuses containing tourists. But with peace and prosperity beginning to spread around the world it was easy to predict the tourist industry and the substantial contribution it could make to government revenue and hence to the well-being of the inhabitants. There were other reasons for establishing national parks, as the United States and a few other countries had already demonstrated—reasons of science, of education, and of aesthetics, as well as of economics. Moreover these two areas were almost uninhabited, owing partly to the evacuation policy in the earlier combat against sleeping sickness.

In the event it was some years before any action took place on this proposal, but Sir Andrew Cohen succeeded Sir John Hall as Governor in 1952 and with his enthusiasm for conservation, not only were the Queen Elizabeth National Park (now Ruwenzori Park) and the Murchison Falls National Park (now Kaberere Park) declared, but also an additional important area in dry country in the far north adjacent to the Sudan, the Kidepo National Park. These completed a trio of some of the finest wildlife areas in Africa. Her Majesty Queen Elizabeth, who was familiar with East Africa's animals, having inherited the throne during the night of 6 February 1952 while staying at Treetops in Kenya, came with Prince Philip to open the national park named after her. It was at the evening party following the ceremony that my colleagues and I were astonished in conversation, to learn how familiar was Prince Philip with East Africa—not only its wildlife but the whole range of scientific, economic, and social problems. The Uganda parks were soon equipped with comfortable lodges and roads and became tourist honeypots until the disaster of Amin's regime. Following his departure and occupation by Tanzanian troops the wildlife and especially the splendid herds of elephant have suffered to such a degree that some observers consider that they are unlikely ever to recover. Optimism tells me that they could, but only if the kind of control which obtained during the early years of independence could be re-introduced.

An integral part of this first development plan for Uganda was concerned with the provision of energy, but this could only be hinted at in the publication of 1946 because it was under negotiation, having wide implications not only for the people and the environment of Uganda but also for those parts of Kenya and Tanganyika adjacent to Lake Victoria, and even more for the Sudan and Egypt which depend on Nile waters. This was the dam and hydroelectric project at the outfall of the Nile from Lake Victoria. Not often are the circumstances leading up to a major scheme of this sort on the record because they so often derive from a number of incidents and ideas which gradually get welded into a whole concept. Therefore it is worth going into a little detail to see just how the waters of the White Nile came to be controlled at their source, by converting Lake Victoria into the biggest reservoir in the world.

Until then Uganda had been almost wholly dependent for energy on firewood, even for conversion to electricity for the larger towns. She was by no means unique in this, for even today it has been calculated that in the world as a whole some 80 per cent of all wood produced is burned as fuel, and with diminishing oil resources, strikes of coal-miners, and a public opinion against nuclear power, more attention is being devoted to wood. In Uganda after the war the railway had been partly adapted to coal and oil, but plantations of gum trees at regular intervals up the line were reminders of the days when frequent fuelling halts were necessary for all trains. But if Uganda was to progress into the twentieth century an additional source of power was desirable, so why not the water power which was so obviously gushing away down the Nile?

Some years before the war there had been a survey of hydroelectric possibilities which had led to various suggestions on a rather small scale, and in 1946 an engineer from England, by name Charles Westlake, had been commissioned to assess the electric demand in Uganda, mainly for domestic purposes in the towns and for minor industrial uses such as ginning cotton. It seemed to me, however, that the provision of electric power should be considered in a longer term, and we should be thinking of demands growing not only during the ten-year planning period but for later decades as well, perhaps even demands outside Uganda which could be supplied by long transmission lines. It was necessary to think big.

Years earlier, when studying fisheries, I had been fascinated by the Ripon Falls (Fig. 20) close to the town of Jinja where the still waters of Lake Victoria plunged northward to form the Nile. Several evenings I went there watching tens of thousands of cormorants on their evening flight from lake to river, and large barbels constantly leaping at the foot of the falls, attempting to reach Lake Victoria. The Ripon Falls themselves involved a vertical drop of only 5 or 6 m, but they were followed by a series of roaring rapids for a mile or so. When sitting there, or casting a spoon into the raging waters to catch a barbel, I was reminded of another prophetic remark of Churchill's when he was at this same spot soon after the turn of the century: 'So much power running to waste . . . such a lever to control the natural forces of Africa ungripped, cannot but vex and stimulate imagination. And what fun to make the immemorial Nile begin its journey by diving through a turbine!'

When working in the Middle East I had become immersed in the problems of Nile control, with H. E. Hurst as mentor. We had discussed the possibility of damming Lake Victoria, thereby providing a vast amount of water storage and starting off the White Nile system with an even flow rather than with fluctuations dependent on periodic and seasonal variations in rainfall. A steady flow at the source would make control by engineering works all the way downstream much easier. When working in

Fig. 20. Ripon Falls before construction of the Owen Falls dam. A barbel (*Barbus altianalis*) is attempting the passage into Lake Victoria.

Uganda, looking down the river from its source rather than up from its mouth, the control of Lake Victoria had even greater attractions. From Uganda's point of view, if a dam was placed a little downstream from the Ripon Falls, to take advantage of the vertical drop of the Owen Falls and the rapids as well as the Ripon Falls themselves, a hydroelectric installation could produce far more electricity than the immediate requirements of Uganda, so there would be surplus for the development of appropriate industries and perhaps for export to Kenya. The main product of Uganda, cotton, was all at that time baled, transported down the railway to Mombasa and thence by ship to Liverpool. After spinning and weaving, a fair proportion of it came back by the same route to clothe the growing populations of Kenya and Uganda. Why not use the local labour and surplus electric energy to convert cotton into cloth on the spot, and save all that transport? Moreover, from the viewpoint of the countries surrounding Lake Victoria, a controlled and predictable water level, even though its range of fluctuations might be increased from the natural 1½ m to perhaps 2½ m, might assist the steamer transport which was difficult at some of the ports at times of low water. An increased rise of water level might cause complications to some of the low-lying shores where there are extensive floating papyrus swamps, but at that time not many people

Fig. 21. Owen Falls dam and hydro-electric installation as seen from the East bank in the engineers' plan.

would be affected. From the viewpoint of Egypt and the Sudan the advantages were very obvious, so I had some private communication with the people in charge of Nile control in those two countries.

The idea was spilled in Uganda first on 29 May 1946, a day for which my diary starts 'A long breakfast with Sir John Hall and Sir Charles Lockhart with the usual discussion about the relations between production and welfare. I mentioned the idea of a dam at Jinja and they both seemed to think that the proposal is a big thing with great promise.' The cost of such a project was clearly well outside the scope of the Uganda Development Plan, but with their encouragement I wrote it up, compressing the many arguments into a couple of pages which were quickly circulated to heads of departments of Uganda Government and others immediately concerned. It caught the imagination and was soon passed with appropriate comments to the Colonial Office in London and to the Governments of Egypt and the Sudan, asking for a conference to be held in Entebbe to consider its implications.

In the meantime, on one of our tours of Uganda, we had called in for lunch at the old Ibis Hotel at Jinja where I met Charles Westlake for the first time. He was very depressed after some weeks of consultation about possible electricity demand. He had met little enthusiasm among the Ugandan authorities and people, African, Indian, or British, and spoke of packing up and returning to his own bailywick in suburban England. I gave him my ideas about a Nile dam with surplus power for industries to process Uganda's crops, and his eyes brightened. He pointed out that almost every assessment of electricity demand throughout the world had been a gross underestimate, for the main demand did not show itself until the power to supply it was available. We went to look at the Ripon Falls together that afternoon.

The conference between Uganda, Sudan, and Egypt happened a little later in the year and its beginning was not auspicious. The group from Cairo was led by the Under Secretary for irrigation and included Dr H. E. Hurst and Dr Amin Pasha, an Egyptian of great bulk who was based in Khartoum in charge of Nile control in the Sudan. The Sudan government sent its head of irrigation backed by several others. The Uganda group made their acquaintance and, on the arrival of the Governor, sat down at a large green baise table. He was in full regalia, cocked hat and all, and opened proceedings with a speech which was comprehensive. He referred to the Nile Waters Agreement of 1929 which reserved the use of Nile waters to Egypt. He explained the needs of Uganda in relation to Nile waters and finished on a slightly beligerent note, saying that Uganda intended to go ahead with hydroelectric production whatever Egypt and the Sudan said. He then stumped out of the room saying, 'Now Dr Worthington will take the chair'.

The rest of that day went very slowly. The possibility of any kind of agreement seemed remote. The advantages of the project were rehearsed time and again, but there was almost no positive reaction. The Nile Waters Agreement was often quoted. I closed the meeting at four in the afternoon in pessimistic mood, and we all resorted to Government House where His Excellency and Lady Hall had invited us to tea.

We sat on the veranda looking across the rose garden to the view of Lake Victoria beyond, and watched the multitude of multicoloured fire-finches taking crumbs off the lawn. Tea finished, the Governor asked 'Anyone for tennis?'. There was a silence, and then his ADC, in duty bound, raised a reluctant hand. No-one else seemed inclined to volunteer so I raised another. Then, to the surprise of all, the enormous Dr Amin thrust his upward, which with the governor himself made the men's four. Amin, being in a dark suit, the ADC was sent to fit him out for tennis. No pair of trousers could be found which would reach round, but the ADC, who was himself large, offered a shirt, and an outsize pair of tennis shoes was found in a cupboard, so all hands resorted to the court. We spun for partners and Amin and I were to oppose the government of Uganda. Amin was to serve and in opening the match he threw the ball immensely high, swiped at it, missed his footing and fell flat on his back. There was a dreadful hush until the ADC leaped over the net to help me in applying first aid. His Excellency called for a doctor, but after a few moments' rest Dr Amin sat up with a broad grin on his face. He then proceeded to play an excellent game to everyone's acclamation.

Next morning the meeting was in good humour, and Dr Amin was elevated to the company of heroes. By lunchtime broad agreement was reached on the principles of the project, and by teatime future actions and the necessary consultations and studies had been sketched out.

Of course there was much yet to be cleared, including a further visit by Egyptian engineers for more detailed consultation. A conference in Nairobi between the governments of Kenya, Uganda, and Tanganyika Territory proved somewhat sticky but eventually objections to the scheme were withdrawn. A prominent hydrologist and civil engineer, Dr C. G. Hawes, was brought over from India to develop a hydrological survey of Lake Victoria and its tributaries. Finance for the project was arranged through His Majesty's Government in London. Leading firms of civil engineers and electrical engineers were appointed as consultants; detailed plans were drawn up, and later came the contracting firms. Meanwhile a Uganda Electricity Board had been established with Charles Westlake as its first chairman. Naturally he was prominent at the moment of climax which came on Thursday 29 April 1954, having been dubbed Sir Charles the day before. On that day the Owen Falls Dam with the first stage of its hydroelectric installation was formally opened by Her Majesty the Queen in the presence of five or six thousand people who were ranged in tiers along the dam and on the opposite side of the river. Shortly before the Queen's arrival the main sluice gates, which were designed to throw water from the foot of the dam in prodigous fountains to dissipate energy before reaching the river below, were closed, so that the Nile was stopped for some 15 minutes. The water level dropped 10 feet or so and there was a mysterious hush, broken only by the flapping of thousands of fishes left high and dry. After the Queen had spoken she opened the sluices with an electric switch and immediately the waters were roaring and the fish swimming once again. There was only one regret: no-one would ever stand again where Speke once stood, and after him Churchill and countless others to gaze at the natural source of the Nile from Lake Victoria, for the Ripon Falls had disappeared beneath the waves.

There is a postscript to this story. As expected the availability of electricity at Jinja led to further demand from many sources: cotton spinning and weaving factories were soon opened at Jinja itself as well as other industries, and a great deal of electricity was sold to Kenya whose own hydroelectric project on the Tana River could not be ready for a number of years. Soon after independence of the country in 1967 and Dr Obote's assumption of the Presidency, it was decided that a second major installation would be needed before long and a survey was made of possible sites. There are several good places: one is not far downstream from the Owen Falls dam itself where a similar vertical drop is available before the Nile flattens and widens into Lake Kioga; another is a series of rapids on the river after it has left Lake Kioga but before it enters the National Park. But the Murchison Falls, where the Nile falls 39 m vertically through a gap in a rocky sill no more than 6 m wide, was the site chosen. This place is the engineer's dream, but it is a very long way from the parts of Uganda where power is needed, and the falls are central to the National Park, a site of pilgrimage for tourists as well as being of high scientific importance in themselves. The disturbance of creating there a major engineering work would be great, so there was an outcry from conservationists, many of whom, led by Sir Julian Huxley, wrote personality to President Obote. I was in an embarrassing position, for having initiated the first hydro-scheme, I found myself among the front runners of those trying to stop the second!

Fig. 22. Murchison Falls on the Victoria Nile, which separates the Nilotic aquatic fauna below from the Victorian above.

President Obote would not be moved, for he had set his heart on this spectacular place. Then came the takeover of Uganda by Sergeant-Major Amin who was keen to do the opposite of Obote, and one of his early acts was to cancel the electricity project at Murchison Falls which he renamed Idi Amin Dada Falls. The conservation lobby was delighted; but this proved to be the only sound act of Amin's reign. Years later Sir Angus Paton then senior partner of Sir Alexander Gibbs and Partners, the engineering firm which had designed the Owen Falls hydro-power works and also the defunct power scheme for the Murchison Falls, assured me that the latter was unlikely to be advanced again. A new study had shown that a site on the Nile further upstream near the Atura rapids and outside the National Park was equally suitable and the second hydro-power station in Uganda would be there.

This first Uganda Development Plan was in some measure a pioneering operation. It had few predecessors in other colonies, and none that appeared to satisfy. It was published and approved by Uganda's Legislative Council in 1946, the year in which it was prepared. The ecological approach to development which it represented was intended as a general guide rather than a detailed programme and I had written in it 'A

Fig. 23. Lake Albert shore: camp below the eastern escarpment of the Uganda Rift Valley after a storm.

development plan which is worth making is worth revising'. It was therefore a source of satisfaction that in 1948, my plan was reprinted with a revision which presented a programme of work and financial provision year by year for the decade. This and subsequent revisions were made under the admirable guidance of Sir Douglas Harris who was appointed Development Commissioner in Uganda following a distinguished career as a civil engineer in India.

The years from 1946 onward saw the blossoming of ideas and action in planning development and welfare in many other colonies in preparation for self-sufficiency and ultimate independence. They also saw the creation of institutes and university departments of development studies. Courses for planners were in vogue, and it was gratifying to learn that this early effort for Uganda, became a text for discussions at lectures and seminars.

In due time, of course, this plan, which ran until 1956, was replaced by a new one, and during the next decade major political events intervened. Sir Andrew Cohen, then as Governor of Uganda, was in the hot seat; he had to take responsibility for expelling the Kabaka, and for a series of other administrative and political measures leading to independence in 1967. But later on, until his untimely death, on the rare occasions that we met, he was wont to remind me of this interesting episode which, though it lasted but a few months, did much to open my eyes to the life and aspirations of developing peoples. Sir John Hall, who was my guide and host in this exercise and wrote an excellent introduction to the Plan, proceeded to a new and distinguished financial career in London until his death in 1979.

7
East Africa (1946–1951)

Lord Hailey in his foreword to Sir Charles Jefferies's review of colonial research (1964) pointed out that the guiding principle in colonial rule throughout the nineteenth and early twentieth centuries was provided by the concept of Trusteeship, which in the last century had received so strong an impulse from the campaign to abolish the slave trade. However this concept was static rather than constructive in its outlook and as a rule involved little in the way of financial sacrifice on the part of the trustee. He went on: It constituted therefore a very substantial departure from previous practice when Parliament passed the Colonial Development and Welfare Act of 1940, which committed Great Britain to make annual grants of considerable magnitude for the explicit purpose of implementing schemes for improving the standard of social and economic life of the colonial peoples.

The 1940 Act had allocated £1 million specifically for research. The original sums were supplemented under successive legislation, so that by 1962 the total expenditure by the British Treasury for the purposes of colonial development and welfare had reached £261 million of which £19 million had been devoted to research.

Scientific assistance to colonial development had, of course, been going on long before the Second World War. It started in 1820 when the government of the day took over the Royal Botanic Gardens at Kew, and from then on Kew did much to assist in agriculture overseas, especially by the exchange and introduction of crops and plantation plants. This help was indispensable, but responsibility for carrying out and paying for research work in the colonial territories remained with the local governments and private enterprise. Then in 1896 a Royal Commission was set up following troubles in the West Indies and this resulted, among other things, in establishing the Imperial Department of Agriculture for the West Indies, at the expense of the United Kingdom. In 1922 this was amalgamated with the Imperial College of Agriculture in Trinidad, which had been opened a year earlier and so then began the great task of training officers for the colonial agricultural service throughout the world, combined with tropical agricultural research.

In medical research there was likewise little organized effort in or on behalf of the colonies until near the end of last century when the scramble for Africa and the 'White Man's Grave' showed the urgency for understanding the causes, methods of prevention, and cure of tropical diseases. In 1899 both the London and the Liverpool Schools of Tropical Medicine were founded with funds provided in the main by private benefactors. Thereafter the Colonial Office did its best to encourage research not only in agriculture and medicine, but also in survey, geology, forestry, entomology, oceanography, and fisheries, but government funds were pitifully small. Most of the knowledge gained about the nature, natural resources, and the peoples of colonial territories continued to depend on private initiatives, enthusiastic amateurs, and expeditions of keen scientists who wanted to work in unknown situations outside Europe.

After the First World War the situation changed with the formation of the Department of Scientific and Industrial Research in 1916, which was interested in the supply of raw materials from the colonies for British industry, and in 1919 a Colonial Research Committee was set up with a vote of £100 000 spread over several years to assist 'less prosperous' territories to conduct necessary research. Meanwhile the Empire Marketing Board was established in 1926 and from then till 1933 spent £285 000 on research in the colonial territories, nearly all on agricultural projects. In 1929 another source of funds for research became available under the Colonial Development Act of that year, which up to 1940 had spent nearly £400 000 on 'research, co-operation, and marketing'.

In 1942 a new Colonial Research Committee was appointed to decide on how the allocation to research under the 1940 Act could best be arranged. This committee was converted into the Colonial Research Council in 1948 and for most of its life Lord Hailey was its chairman. At the end of the war his high reputation as an Africanist had been much enhanced by having, amongst other activities, resided in the Belgian Congo for a while to hold that enormous and rich country firmly on the allied side while Belgium was under German occupation.

As wartime restrictions ended it became possible to spend the funds allocated to colonial research, but they could hardly stretch to provide an effective service for all the colonies, protectorates, and mandates. Would it be possible, I thought, to break with tradition and, instead of spreading the butter thinly on all the pieces of bread, to put a reasonable dollop on one or two, even at the expense of a reduction of the rest. Thereby one might in one region consisting of a group of colonial territories establish a series of research organizations to serve several countries at the same time, with a better chance of having a real impact on the problems of development and welfare.

At that time I had been looking forward to a period of reconstruction of the Freshwater Biological Association in a new headquarters more convenient than Wray Castle, and, with my family, to enjoy the amenities of country life in the Lake District freed from wartime restraints. This was an attractive prospect, but I could not refrain from writing a one-page note to Lord Hailey expressing views on colonial research. He read it to his Council and I received a summons to attend its next meeting. The Council was at that time keen to appoint a scientist with overseas experience as its secretary. They offered me the job, and after some deep cogitation and with some reluctance I accepted, but on the understanding that my work would not be based in Whitehall. I firmly believed that to organize colonial research effectively one needed to live and work in the colonies in close touch with the people who were on the job. I suggested that the first main effort be made in East Africa to serve the four territories of Kenya, Uganda, Tanganyika Territory, and Zanzibar.

For years there had been a good deal of co-operation between these countries through the East African Governor's Conference which, after the war, was in process of being converted into the East Africa High Commission, to be based in Nairobi and to operate with a high-powered Administrator, Sir George Sandford. He would be responsible for the pre-existing Services common to the East African region and any new ones which might be established. Andrew Cohen, who at the time was in charge of the East African desk at the Colonial Office, suggested that, on appointment as Secretary to the Colonial Research Council, I should be seconded immediately as Scientific Secretary to the East Africa High Commission. This suited me and it was arranged that I would go to Nairobi for several years to work up science in the East African region with the full support of the Colonial Office. Meanwhile the day-to-day administration of colonial research elsewhere would be looked after by others in Whitehall. To my wife and family, if they had to leave the Lake District, Nairobi sounded more attractive than London for we still had friends out there from the African expeditions of 1927–31. However, their move was deferred for some months owing to the episode in Uganda which is recounted in the previous chapter.

Once established in Nairobi the initial task was to prepare a sort of grand design for research services in the four countries. It had to be adjusted to the total ecology of the region, the economic and social development plans for each country covering the coming decade, and their administrative and political patterns. But in order to design the future, past events which led to the current situation had to be reviewed.

Moves towards association of the British East African countries had started in 1924 with a Commission headed by Ormsby-Gore (Lord Harlech), then Secretary of State for Colonies. Each country was too small to be economically viable on its own and a pooling of resources could lead to greater efficiency in administration and development. These ideas had been advanced by the Hilton Young Commission of 1928 with which as a very young man, I had discussed fisheries at Entebbe during my first expedition. Sir Hilton Young had laid special emphasis on the need for co-operation in research and science. On a broader scale the Hailey Survey of 1934–7 pointed to the same principle, but also, by drawing comparison between one region and another, showed the need for closer co-operation between the European powers responsible for the different parts of Africa.

In East Africa, up to the outbreak of war in 1939, contact was maintained in the fields of medicine, veterinary work, various branches of agriculture and the like, by periodic discussions between directors of technical departments and by meetings of specialists. Although the railways and harbours administration, telecommunications, customs and excise, and other forms of taxation were already run in common, the only scientific service with truly inter-territorial functions was the East African meteorological department which was created in 1929 in order to collect under one control a series of stations along the Imperial Airways routes. For some branches of agriculture, however, there was the Research Institute at Amani in Tanganyika Territory which had been established a year or two earlier under the British Mandate on the foundations laid by the Germans before the First World War. In addition the Tsetse Department of Tanganyika and the Veterinary Department of Kenya were, by agreement, performing certain functions of use in all the territories.

Not many branches of science which are intended to advance economic and social progress proceed well in isolation, and the isolation of individual workers in the separate territorial technical departments, especially those who were posted to outlying field-stations and hospitals, had been increased during the war. This was recognized at various discussions which had already taken place on the enlargement of research services, made possible by the CD&W funds but in my view there had been too much focus on the form of central organization and too little on the men who were to conduct the practical work. Before my arrival there had already been several visits by scientific specialists from the United Kingdom to prepare the way in establishing inter-territorial research and we continued such arrangements. The visitor studied the position in East Africa and made proposals which were considered before his departure by a conference of territorial representatives and specialists. This led to applications for grant aid to be voted in Whitehall.

In all these matters support from the top, that is from the officers heading the administrations in the four component countries of East Africa, was all important. Sir John Hall was Governor of Uganda and always ready to support my views after our association over the Development Plan for his country. Sir Edward Twining, Governor of Tanganyika, was ready to look at new ideas and understood the value of science, and Sir Vincent Glenday, resident in Zanzibar, was an old friend from the time when he was Provincial Commissioner in northern Kenya and I was conducting expeditions to the great lakes. The ex officio Chairman of the East Africa High Commission was, however, Sir Philip Mitchell, Governor of Kenya, who was apt to take a somewhat jaundiced view of scientists. During one of my few talks with him he asked for examples of where research had truly advanced development in East Africa. I gave him three: the analysis by Amani scientists of the soil catina which had greatly assisted practical work in controlling soil erosion, research at his own veterinary department at Kabete leading to the immunization of cattle against rindepest over a large part of tropical Africa, and the ecological study of Lake George in Uganda which had resulted in a flourishing commercial fishery. He remained unimpressed and seemingly unhelpful, but did not, I hasten to add, do anything to impede development of the inter-territorial research services. Fortunately my dealings with the Chairman were generally through Sir George Sandford who, as Administrator, had wide understanding, quick wit, and absolute integrity. He proceeded later to Governorship of Bermuda where he died tragically during his first tour of duty.

At an early stage there was occasionally a problem in convincing the directors of technical departments in the territories that the establishment of inter-territorial research organizations would not reduce their own responsibilities or take from them some of the more active scientific units under their command. Our intentions were of course quite different, namely to add to the pre-existing pattern a number of institutes which would undertake research of the more fundamental kind, though still related to practical needs. Such research, some of long term, could not be done effectively by territorial departments owing to their pressing day-to-day needs. These regional institutes would be based at strategic centres in any of the four territories, but some of their staff and facilities would be posted at the sites of particular problems under investigation, where they would enhance the work of territorial departments.

The critics would argue that the CD&W funds would be better spent if handed direct to the territorial departments to help meet their local needs. Perhaps they had a point because, when political independence came in the 1960s, much earlier than most people had anticipated, territorial rivalries and misunderstandings led to the breakdown of a good deal that we had established under the East Africa High Commission, by then called The East Africa Community.

However, looking forward in 1946, there was a good spell of years ahead and the path was clear: East Africa became a guinea-pig for the establishment of overseas regional research and took a large slice of the financial cake. Later a similar system was created in the West African group of British territories—Nigeria, Ghana, Sierra Leone, and Gambia, but there were problems rising from the lack of contiguity of the four countries, divided as they were by wedges of French administration. In what was then known as Central Africa, namely Northern and Southern Rhodesia and Nyasaland, interterritorial research, particularly in agricultural subjects, also became important. But Central Africa's scientific co-operation was short lived owing to well-known and drastic political changes.

In the design of research services for East Africa we took note of the fact that most government-sponsored science up to that time had been under the thumb of administrators. Research, if undertaken at all, was focused on local problems, and there were very few groups of research workers who could generate their own programmes and ideas: 'The place of the scientist is on tap but not on top' was the rule and this tended to stifle bright young scientists.

For this reason we separated research from administrative responsibility and gave the scientists a rather free hand in proposing and carrying out their programmes. Much of the research was arranged in a series of organizations with functions extending over the whole region and these organizations undertook research projects on problems which had wide significance. The regional research organizations served in an advisory capacity to territorial departments concerned with agriculture, veterinary work, forestry, fisheries, tsetse control, health and medicine, and public works but decisions on practical applications remained a territorial, not a regional, responsibility. Scientists in the regional organizations maintained close touch with their opposite numbers in territorial departments through visits and frequent technical meetings.

There were four substantial advantages in pooling a good part of the new research effort in this way: first, greater efficiency, including savings in capital expenditure and personnel; second, better recruiting prospects, because regional research organizations offered wider scope to young scientists; third, easier arrangements for seconding specialists to and from centres of research in Great Britain, the Dominions and other countries overseas; and fourth, better service to the territorial governments and those interested in using the results of research. We were in fact following the well-proven system of the Research Councils of the United Kingdom which, with the institutes which they operated or grant-aided, had a high measure of autonomy at that time and had been successful in separating research from executive control. It was a system quite different from the 'Rothschild Principle' which was adopted for government-sponsored research in the United Kingdom during the 1960s (see Chapter 11). By that time some British research undertaken with government finance had perhaps drifted too far away from practical needs and Lord Rothschild's recommendations helped to redress the balance. In East Africa, however, there is no question that the research organizations' relative autonomy, on which we insisted, gave a new stimulus to the research workers. Research was kept in line with practical needs through a system of committees composed of directors of the relevant territorial departments with the regional organization's director in the chair. .

The first six regional research organizations to be established were for the following subjects: agriculture and forestry, which was based on a new institute at Muguga some 17 miles from Nairobi in Kenya; veterinary science and animal husbandry, which was based in a separate set of buildings also at Muguga with its own field facilities; tsetse flies and trypanosomiasis, which was located at Tororo in Uganda near the Kenya border, but with the former headquarters at Shinyanga in Tanganyika Territory maintained as an outstation; inland fisheries, which was based at Jinja in Uganda on the northern shore of Lake Victoria; marine fisheries, established at Zanzibar; industrial research, based in Nairobi. In the medical field the Virus Research Institute at Entebbe in Uganda was an important element in the overall pattern; it was developed from the former Yellow Fever Research laboratory of the Rockefeller Foundation and achieved much fundamental work on the viruses themselves, the wild fauna in which they subsist and the insects which transmit them from host to host. Regional malarial work was re-posted to Amani in Tanganyika and there were several other units for health and medical research, the largest of which was based at Mwanza in Tanganyika. Also in Tanganyika a centre for insecticide studies was created near Arusha not far from the Kenya border. An institute for sociological research was attached to Makerere College in Uganda, while the headquarters of the East Africa High Commission in Nairobi incuded economics and statistical offices.

Research studies undertaken in a number of other subjects also came into the total picture and periodic co-ordinating conferences, rotating around the territories, were instituted at this time. This applied for instance to geology, for which each country had its own active survey; hydrology which was becoming recognized as one of the most important sciences needing development; locusts in which the East African studies were related to those based in London and later to FAO in Rome; and wild fauna and its conservation in National Parks and Reserves. An East African Herbarium was established to incorporate several important collections in the territories, and archaeological research undertaken by Louis Leakey and his collaborators based on the Coryndon (later National) Museum in Nairobi was related to the rest.

At this time there was only one major centre for higher learning in East Africa, namely Makerere College at Kampala in Uganda which was soon converted into the first University in East Africa and became an important centre for research. A technical college was being developed in

Nairobi and later became the nucleus for the second university in the region, with that at Dar-es-Salaam to be inaugurated some years later. Of other non-government scientific activities the East African Natural History Society was already of long standing, had published extensively, and had an important library situated at the Coryndon Museum. As a second forum for the presentation and discussion of research and topics of general interest we established a Nairobi Scientific and Philosophic Society.

In order to give some general impression of what we actually achieved in East Africa during these five formative years let us glance at the subjects in the ecological sequence, starting with the shape of the land and finishing with the human population.

Survey and mapping are basic to all forms of understanding and development and in 1947 Britain had created its Central Directorate of colonial geodetic and topographic surveys. The geodetic framework of Africa was more or less adequate by then so their main effort was on air survey, country by country, to produce large-scale topographic maps as an aid to all kinds of development. The preliminary plots from air photographs on scale 1:50 000, which were circulated for checking of place names and ground features, were most useful to men on the spot. Co-ordination was done in London leaving the territorial departments of lands and surveys to concentrate mainly on cadastral work. There was however one important aspect of mapping which still needed stimulus, namely maps to illustrate particular distributions, such as soils and vegetation, agriculture and animal industry, climate, tribes, and population. Such maps, ideally in the form of transparencies which could be placed one on another, were the essence of ecological planning and, even if reduced to the scale of a school atlas, were of high value, not only for education. Of the four countries concerned Tanganyika Territory was at that time the only one to have such an atlas.

Geology provided a special case of mapping which helped greatly in understanding physical features and soils and was essential for any form of mineral industry, even if concerned only with road metal and building materials. Members of each territorial geological survey, together with administrative officers, gave life to the early topographic maps which, in the absence of obvious features such as hills or rivers had large blank areas. One could appreciate the feelings of early surveyors before the days of air photography, armed with plane table and level, who sometimes scrawled MMBA across a sheet, which being interpreted, indicated 'miles and miles of bloody Africa'. The territorial geological surveys added incident; they were progressing well but were independent one from another, so the case for some kind of closer collaboration between them was strong. The possibility of amalgamating these geological surveys into one for the region had been considered and discarded some years earlier so we focused attention on co-operation and this led to periodic geological conferences and, among other things, to the first geological map covering East Africa as a whole.

Meteorology, already a regional service, was expanding to meet the growing needs of air transport. The service was also one of the most essential in any development concerned with land and water use, so the network of climatic recording stations was expanded. It should be noted in this connection that when the East African ground-nut scheme was under consideration soon after the war advice was solicited and freely given from the regional scientific services such as they then existed, as well as from the territorial departments. Although there were no climatic records then available close to Kongwa in Tanganyika and the other areas selected for the ground-nut scheme, the predictions of rainfall and other climatic factors, as extrapolated from existing stations, indicated that the chances of growing ground-nuts on a large scale in the selected areas were poor. Similar advice came from other local considerations, but in the political climate of the time it was overruled in London. The ecological picture was fairly obvious even at that time: the areas selected for the ground-nut scheme were almost entirely unpopulated because the particular conditions of climate, soil, vegetation, and animal life were not conducive to agriculture. Bulldozing the vegetation and adding fertilizer to the soil did not ameliorate the climate!

Hydrology at that time fell largely into the province of geological deply. Accordingly the High Commission initiated a series of hydrological conferences bringing together the geologists, engineers, agriculturalists, were very few records of the flow in rivers and streams. Fluctuations in flow, annually and over a period of years, are of the very essence when it comes to designing engineering works for water control and water sup-Accordingly the High Commission initiated a series of hydrological conferences bringing together the geologists, engineers, agriculturalists, and and health authorities, all of whom were deeply concerned with water from one viewpoint or another, reinforced by the few hydrologists who were already working in the region.

Background biology, including the production of reference works on flora and fauna, had been filled in to some extent by enthusiastic workers, some professional and others amateur, so there were already a number of books concerning the more obvious animals like mammals and birds, and one or two volumes covering the great variety of indigenous trees had been produced by enthusiastic forest officers. There had never been, however, in Africa, anything approaching the surveys of flora and fauna such as had been organized for the Indian sub-continent. In consequence it was often impossible even to put a name to common plants or insects without reference to leading herbaria or museums in Europe or America, and even then a good many forms proved to be new to science. Therefore, in close co-operation with Kew Gardens, we stimulated the production of a Flora of East Africa and established what has become a comprehensive herbarium for the region in Nairobi. Even now the *Flora*, which runs into many volumes, is not complete, but it provides a reference work which will last for ever. We had plans for some similar coverage for the equally diverse fauna to include insects and other invertebrate groups, but this project unfortunately never got beyond the drawing-board stage. Nevertheless the Coryndon Memorial Museum had become one of the best in any colony, built up by voluntary effort with the aid of small grants from Kenya Government and Nairobi municipality. During these years a special subscription list allowed the building of an extension and the enthusiastic but underpaid staff gained wide recognition.

Not only in the classical studies of systematic botany and zoology were advances made, but there was a small but growing number of biologists who were more interested in what plants and animals did than what they were. Thus the disciplines of ecology and ethology were getting established. Some people in authority began to realize that the principles of ecology could be applied with advantage far beyond the wild plants and animals which had provided the material for its original thinking.

The Great Game animals for which East Africa was famous warranted special consideration owing to their importance to the economy as well as to science and to ethics. On one hand some species caused serious damage to crops and, being themselves immune, were agents in spreading local diseases to domestic animals. On the other hand their products, especially of ivory, rhino-horn, and shooting licences, produced significant revenue. Of great importance to the welfare of big game as well as of birds was the growing tourist industry, in which the field-glass and camera were rapidly replacing rifle and shotgun. Provided sufficient of the wild fauna survived in accessible areas tourism gave promise of becoming a major source of revenue.

Even by 1946 the wild fauna was seriously reduced in numbers and area of occupation. The territorial game departments had the double function of controlling the damage caused by wild animals and of conserving the resource. Fortunately these departments were for the most part staffed by officers of high responsibility, who had designated, looked after, and saved from 'development' some of the best wildlife areas as game reserves. Soon the interest in wildlife and wild places had reached the stage of converting the best of these areas into National Parks under authorities separate from game departments. In later years some writers about wildlife in Africa have given the impression that game departments existed to destroy wildlife, national parks to save it. Nothing could be further from the truth, for the Game Wardens were among the first and most pressing of those to advocate national parks. When, for example, Captain Archie Ritchie, the chief game warden of Kenya, backed by his many friends, had succeeded in this, the government asked him to take charge of the parks as well as the game department. He said, 'No. You must create a separate authority with a director, separate from government. That is the way to defend the parks from the pressures for land which will mount against them.' Had it not been for the continuing work of game departments over some 30 years there would have been few animals left to provide a reason for national parks.

However, both the game departments and embryonic national parks authorities in each country needed guidance in their management programmes. Questions were frequently being asked such as: What species of wild animals suffer from or are carriers of the diseases of domestic animals? What areas of land do the different species require for their health and happiness? What are their feeding habits, breeding rates, and other relations to the environment? What are their movements and migrations in the different areas and how can they be determined accurately and if necessary controlled? How can land, left mainly to nature, be managed in order to provide its maximum carrying capacity for wild life and perhaps a sustained yield of venison? In order to stimulate interest and provide guidance in answering questions of this sort we instituted a series of periodic, wildlife conferences for the exchange of information and advice. But unlike the other subjects we went wider than the East African territories to include also experience from Central Africa—Northern and Southern Rhodesia and Nyasaland.

Agriculture and forestry, Animal health, and *Tsetse flies* each formed the subject of a new research organization established under the East Africa High Commission. They were in fact among the largest of these entities, in staff, expenditure, and scientific output and continued to expand and serve the region up to the time of political independence. When political differences between the countries made co-operation almost impossible, even among scientists, the regional organization ceased to function as such. The institutes continued but with responsibility to the territory in which they happened to be situated rather than to the region. As envisaged from the beginning they served an important function in scientific training and the proportion of indigenous to expatriate staff has increased steadily. The literature on agriculture and related subjects in Africa is already so extensive that one hesitates to write more, but it is worth noting that some excellent scientific reputations were made during these years by a number of the staff scientists, particularly in such subjects as plant virus diseases, relations between soil, hydrology, and plant cover, and the ecological complex of tsetse flies, trypanosomes, wild and domestic animals, and man. Publications which record such work illustrate well the changing attitudes to research in the tropics around the middle of the ecological century and in this connection it is noteworthy that, following a successful ecological survey of Northern Rhodesia (Trapnell 1937) which had become accepted as the guide to land use in that country, its author was appointed to the East African Agricultural and Forestry Research Organization in order to inaugurate a training scheme for plant ecologists who would undertake similar work elsewhere.

Locusts have been major pests of agriculture in Africa since at least the time of the Old Testament. Of the three important species the migratory locust affected mainly western Africa, although it had on occasion extended its range to most of the continent; its permanent breeding sites are near the Sahara. The red locust affected southern Africa with permanent breeding sites in the neighbourhood of Lake Rukwa. It was primarily the desert locust (*Schistocirca gregaria*) with which we were concerned in East Africa. B. P. Uvarov (later Sir Boris) had discovered that this insect, like the other locusts, has two phases, one solitary, confined mainly to marshy areas in the Arabian peninsula, the other migratory. The migratory phase, which is distinguished morphologically as well as in its behaviour, appears when the solitary phase builds up large populations in confined habitats. It then forms swarms which used to fly over nearly the whole of eastern Africa spreading destruction wherever they settled down to feed and breed. The fact that locust invasions have in most cases now been brought under control is one of the great achievements of applied ecology. It has involved a great deal of botanical and zoological and climatic research and a permanent organization of staff and transport by land and air for distributing toxic chemicals at the right place and time. An important step was establishment of the Desert Locust Survey as a department of the East Africa High Commission in 1948. This was linked to the international organization for locust control based on FAO in Rome.

Related to this was the *Colonial Insecticides Unit* which was established near Arusha in Tanganyika, a strategic base from which research could be conducted on agricultural, medical, and veterinary problems in a wide variety of habitats. Its purview was extended to fungicides and herbicides but the main object was study of the effects on mosquitoes and tsetse flies and on the environment of the new insecticides such as DDT and BHC in different formulations and under different methods of application. This active unit was responsible direct to the Colonial Office in London because it served all the colonies, but being in East Africa it contributed an important part of the regional research effort, especially in its studies of plant and animal ecology which were necessary to assess the effects of toxic chemicals.

Fisheries are apt to be regarded as the cinderella of public services; but in Africa they are very important as providers of animal protein to the diet of millions of people who would otherwise be almost without it owing to the diseases of domestic animals which cause wide areas to be empty of meat. Around the coasts sea fish was generally available, though the seas bordering the east coast of Africa are by no means as productive as the west. Away from the coastal belt sea fish rarely penetrated inland except for luxury markets, but inland fisheries on the great lakes and rivers provided a lot of food locally, and dried fish had from time immemorial been an important item of internal commerce.

During twenty years since the original fishery surveys (Chapter 2) inland fisheries in East Africa had expanded considerably. Since fish are mostly wild fauna they had been looked after by game departments, but their importance was such that separate fishery departments, or sections, were already being advocated, especially after the appointment of C. F. Hickling as fisheries adviser at the Colonial Office in London. The recommendation which we had made after the fisheries survey of Lake Victoria that a permanent institute of inland fisheries be established in the region had not yet been implemented, but the inauguration of the High Commission indicated that the time was now right. Indeed I had started preliminary moves during the development planning of Uganda, with Jinja at the source of the Nile being earmarked as the most appropriate centre. To save time and money I had selected one of the more successful designs for laboratories, that of the Virus Research Institute at Entebbe, and as soon as a grant was available from CD&W funds, a laboratory of very similar proportions was built close to the landing stage at Jinja, with staff housing adjacent. Thus inland fisheries became the subject of the first East African research organization to be established. R. S. A. Beauchamp, who had inaugurated research of the Freshwater Biological Association in Britain and had subsequently conducted limnological research on Lakes Tanganyika and Nyasa, was appointed director. He set a high scientific standard and attracted a succession of outstanding men and women to his staff over a period of 13 years. They advanced knowledge of Lake Victoria and its fisheries to a great extent and inaugurated a number of researches further afield in other waters of the region.

However, research alone was not enough to improve fishery production and management, so we set up also under the High Commission a Lake Victoria Fisheries Service, supervised by a Board on which sat the three Provincial Commissioners, one each from Kenya, Uganda, and Tanganyika, who were responsible for administration of the lake's shoreline, together with the director of the Jinja laboratory, and myself as chairman. We were lucky to appoint as chief fisheries officer George Cole, a marine officer who had spent much of his career surveying the coast and inshore seas of Tanganyika Territory. In his new responsibility he appointed a fisheries officer in each territorial section of the lake, and each of them was equipped with a sea-going well equipped fishery launch and a number of centres ashore and a seagoing vessel. The fishery was thus brought under a system of scientific co-ordination and control which was much needed. Unfortunately this could not outlast independence, when each country went its separate way. George Cole, after retirement, achieved fame with his wife and 'granny' aged 90, for an epic voyage in a home-built catamaran from Mombasa, via the Seychelles, the Indian Ocean, and Torres Straits, to Australia, and then on to New Zealand.

In the pattern of research organizations sea fisheries were added later by establishing an institute at Zanzibar, which also did excellent work, up and down the coastline of Kenya and Tanganyika, and further afield in the Indian Ocean around the Seychelles. Zanzibar became the home also of another research activity attached to the EAHC, namely an intensive study of a virus disease of clove trees which were and still are a major source of that island's prosperity. For this work the historic house once occupied by Livingstone provided headquarters.

We could go on to relate scientific activity in other branches of the EAHC, such as the use of local materials for purposes of construction and road-making by the EA Industrial research organization, and human sciences including nutrition, medicine, and sociology. But enough has been written to indicate what we were after, namely provision within the countries of East Africa of an infrastructure of research, including the training of local staff, so that knowledge about the lands and waters, their natural resources and their peoples could provide a sound basis for economic and social progress and all the decisions about development which would soon have to be made.

Some might think that the efforts of the British, and also the French, Belgian, and Portuguese, to put colonial science on its feet during the last decade or so of the colonial era was abortive, because in the forms described in this and the following chapter they could not survive the winds of change which blew through Africa in the 1960s. Although technical assistance from Europe and elsewhere continued, and indeed was enhanced from the United Nations, the USA, and other advanced countries, a good deal of disruption was inevitable under independent African governments. The moves towards closer union between groups of territories which were desirable because few single territories were economically viable on their own, went into reverse on independence so that regional organizations could no longer exercise their designed functions. Few of the experienced scientists who knew the problems were prepared to continue under inefficient organizations and so were replaced, mainly by expatriates from non-colonial countries to whom the local problems were strange. Some indigenous scientists who had received training certainly came into their own, but a few others who had yet to acquire strict scientific discipline, did little to enhance the reputations of the institutes in which they worked during the years immediately after independence.

Concerning expatriate scientists, a number of research workers came from the USA on Fulbright fellowships during the effective years of the East Africa High Commission and our experience of them was somewhat varied. The senior fellows, men and women established in their profession, were of great value almost without exception. They spent a few months with us, introduced new concepts and sometimes initiated useful scientific projects. Some of the junior fellows, who came to Africa soon after qualification, were also outstanding, and proceeded from strength to strength in their chosen fields. But their tours were limited to three years or less of which the first was spent getting used to the environment, and the second learning the job, so that their input to African problems was limited. Not many came back for a second tour when they could have done work of great value. The majority acquired good experience but did not contribute as much as had been hoped.

In spite of the rather short effective life of the regional scientific services they laid a tradition in the quality of East African research which still stands in good stead. The institutes we built up have survived. New centres of research and new traditions have been added to the old, and who knows what will happen when the countries of East Africa come once again to collaborate with each other? Meanwhile I pay tribute to many colleagues who built up and worked in these organizations, men and women such as Sir Bernard Keen, H. H. Storey, Sir Charles Pereira in agriculture, Howard Binns in veterinary science, R. S. A. Beauchamp in hydrobiology, D. Baxter Wilson in malariology, K. S. Hocking in insecticides, A. C. Rainey in locust research, Audrey Richards in sociology, four of whom have been elected to fellowship of the Royal Society. Our endeavours were supported by many organizations connected with the Colonial Office in London, especially the Colonial Research Council, and the Advisory Committees on anti-locust research; agriculture, animal health and forestry research; fisheries; insecticides, colonial medical research; colonial products research; social science research; tsetse and trypanosomiasis research. Together Council and Committees brought the experience of more than 250 of the leading men and women in the United Kingdom who were concerned in these subjects, to bear on colonial problems. From a personal point of view this time of scientific administration greatly enhanced my appreciation of the 'holism' of ecology, and indicated at the very middle of the century that an environmental revolution was on the way.

These years based in East Africa provided many interesting adventures and perhaps I may record one or two in which my family were involved.

Like most British in Africa we delighted in viewing big game, and over a New Year holiday were travelling through Masailand *en route* to an assignation with the game warden in charge of the Serengeti Reserve, as it was before it became a National Park. The weather was fair and as night came we put up camp beds in the bush not bothering with tents. I was up early next morning taking a walk with the intent of adding a guineafowl to the larder, and returned to find wife and daughters still in bed, each surrounded by a group of Masai Moran dressed in nothing but red-ochre and blanket flapping in the breeze, their great spears in hand, leaning over the beds to examine these peculiar creatures. The girls, very wide awake and anxious to rise and dress, were trying to communicate with them. Masai Moran, the warrior class, live together under strict discipline for several years after initiation: with the situation once appreciated, they withdrew with excellent grace.

In those days of leisurely travel, there was opportunity to see and do things *en route*. On one occasion soon after the war when we were homeward bound by sea, the Sudan Government were worried about a young German to whom they had issued a short-term permit to study coral reefs at Port Sudan. He had applied to extend his time there and perhaps to set up a base for aqualung diving, which was then in its infancy. A senior administrator in Khartoum whom I had met in the days of MESC asked me, while our ship was at Port Sudan, to check whether this German was a serious biologist, or was, as they rather suspected, some kind of post-war German agent. We found him a very charming man, with a couple of diver-biologists also from Germany and a girl with flowing blond hair who looked superb when swimming underwater among the corals and multi-coloured fishes. We went out with them, ourselves lying on the surface with water goggles, and watched the divers working with aqualungs fathoms below in these exceptionally clear waters. We discussed studies on marine life, including the rate of growth of corals as shown on several old wrecks whose dates of sinking were on the record. Our German friend was Hans Haas, a pioneer of what has since become a common technique in marine ecology and a fascinating hobby as well. I put in a good report. Hans Haas went on to become famous for his undersea films, books and broadcasts—and to marry his photogenic girlfriend.

On another occasion we broke a journey in Italy for a short cultural tour. It was in February, very cold and, having not yet recovered from wartime shortages, the Ufizzi Gallery in Florence had no heating. My youngest daughter, bored with the pictures, ejaculated in Kiswahili 'Mimi baridi sana' (I'm very cold). She was answered by a uniformed guard 'Jambo, Memsahib kidogo, moto iko' (Hello, little Madam, here's warmth). He led us to his little room where there was a wood fire, and when we had warmed up he took us on a tour of the gallery during which we discussed Italian art in 'kitchen' Kiswahili. He had been a prisoner from the Ethiopian campaign and had spent much of the war working on a settler's farm in Kenya, to which he was only too keen to return.

8
Pan-African science (1951–1956)

After the Second World War new ideas on colonial development, welfare, and science were by no means limited to Britain and her overseas territories. Lord Hailey's African Survey had stimulated thoughts about the advantages of closer co-operation between the countries, especially in science and technology, so moves in that direction began soon after the territorial governments had got reconstructed on a peace-time basis.

At that time Africa south of the Sahara could be envisaged as a number of regions although not all of them had any kind of regional administration. Four of these regions had English as their dominant language and contained more than half the total population: two of them, British East Africa and British West Africa have been discussed in the last chapter. The third used to be referred to as Central Africa and later became the Federation of Rhodesia and Nyasaland which had but a short life: it consisted of Northern Rhodesia (now Zambia), Nyasaland (now Malawi), and Rhodesia (now Zimbabwe). The fourth was the Union of South Africa together with the mandated territory of South-West Africa and the British Protectorates of Basutoland, Swaziland, and Bechuanaland (now Botswana).

Four other regions had French as the dominant language, three being the responsibility of France and one of Belgium. Afrique Occidentale Française (AOF) included Sénégal, Mauritanie, Guinée, Côte d'Ivoire, Soudan, Haute Volta, Niger and Dahomey, and the mandated territory of Togo Française. Afrique Equatoriale Française (AEF) comprised Moyen Congo, Gabon, Ubangi-Shari and Tshad, with the mandate of Camerun-Français attached. Madagascar, with which it was convenient to include the islands of Réunion, Comores, and other Mascarene islands, was included in the African concept although it had many unique features, both natural and ethnic. The fourth francophone region was the Congo Belge together with the mandated Ruanda-Urundi.

Two Overseas Provinces of Portugal — Angola, and Moçambique, being widely separate from each other, were best considered as separate regions, whereas Guiné Portuguesa was more appropriately regarded as an enclave in the AOF. The Portuguese islands of Sao Tomé and Principe had closer affinities with the AEF. Finally there was the Spanish island of Fernando Po with its mainland appendage of Rio Muni, but they, together with the Sudan, Ethiopia, the Somalilands, and Seychelles remained on the fringe rather than as an integral part of Pan-African science.

All the countries individually, and in the British, French, and Belgian regions to some extent collectively, supported their own technical services of agriculture, health, education, public works, and so forth. In the larger countries the government departments concerned included scientists who conducted research on the country's natural resources and people, when not engaged on routine duties. In addition, and of greater significance when it came to international co-operation, the regions were actively developing research institutions which were broadly comparable to those described for British East Africa in the last chapter. Thus in the AOF, extended later to the AEF, there was established before the war the Institut Français d'Afrique Noire (IFAN) under the leadership of that great French biologist Theodore Monod. Now at the Musée d'Histoire Naturelle in Paris, he is an ascetic ecologist–philosopher with a powerful sense of humour; he was never so happy as on a camel in the central Sahara with nothing but a pocket full of dates. Monod placed IFAN's headquarters at Dakar and established branches specializing in different disciplines in most of the countries of AOF and AEF.

After the war the French Government created its Office de Recherche Scientifique et Technique d'Outre Mer (ORSTOM)—the T was added to its original title in 1955. Under that organization regional multidisciplinary research institutes were set up in AOF, AEF, Madagascar, and Camerun, all of which grew into important centres for African studies. Even now a decade or so after independence, ORSTOM continues to service the former French colonies with scientific staff and facilities.

The Institut National pour l'Etude Agronomique du Congo Belge (INEAC) was created by Royal Decree in 1933. After the war INEAC grew to very large dimensions with headquarters at Stanleyville (now Kisangani) and many substations for research in agriculture, animal husbandry, and related sciences. The eminent plant ecologist J. Lebrun who had conducted much field-work in the Congo was deeply concerned in its activities. The Belgian Institut pour la Recherche Scientifique en Afrique Centrale (IRSAC) was founded after the war for subjects other than those falling under the broad head of agriculture. It was designed to have a fundamental and in some degree international programme of research. Under the directorship of Louis van den Berg, whose early work had been in tropical medicine based at Antwerp, an impressive multidisciplinary headquarters was built on the shores of Lake Kivu near Bukavu with several out-stations including one for hydrobiology on Lake Tanganyika, another for anthropology in Ruanda-Urundi, and a third on Lake Tumba in the great lowland Ituri forest. IRSAC had a big influence on African science until the Belgian administration was withdrawn. Its

Fig. 24. Regions of Africa south of the Sahara for the purposes of CSA/CCTA.

research then became almost impossible because IRSAC's main centres were in areas of strife where law and order ceased to exist for several years.

In Portugal a Council for Scientific Studies in the Overseas Provinces was likewise established soon after the war and set about creating a number of new research centres in Angola and Moçambique.

Compared with the other African regions the Union of South Africa was far advanced in its scientific activities, for it had a thirty-year start with an economy which could afford a lot of official, university, and private research. Much new activity was grouped in a number of institutes under its Council for Scientific and Industrial Research (CSIR) which was set up after the war. Next to South Africa Rhodesia was at that time the country best endowed with scientific potential for developing its natural resources.

Early moves towards international collaboration were made by the scientists themselves and were expressed at an Empire Scientific Conference convened by the Royal Society of London in 1946. Basil Schonland, whom I had known at Caius College, Cambridge, when he was a research fellow and I an undergraduate, took a prominent part. He was by then an eminent atomic physicist, close friend of General Smuts and had been appointed chairman of the CSIR. It was arranged that South Africa should convene a major African scientific conference in Johannesburg and this took place in 1949.

On that occasion representative scientists from nearly every country in Africa south of the Sahara reviewed the whole field and laid out an initial design for scientific collaboration. It led to one of the more remarkable enterprises in bringing the sciences, with ecology in its broad sense in a central position, into the service of mankind. This enterprise expanded during the following decade until the winds of change led to independence of the separate countries, and the regional concept then lost its roots.

The fundamental recommendation of the 1949 conference, which was quickly formalized by the six governments concerned (Belgium, France, Portugal, Rhodesia, Union of South Africa, and United Kingdom), was to set up a Scientific Council for Africa South of the Sahara (CSA as derived from the French title, Conseil Scientifique pour l'Afrique). This was to be composed entirely of scientists drawn from the member governments and mostly working in Africa, with a Secretary-General and staff who would live in Africa and travel widely to keep in touch with scientific institutes and individuals. Out of CSA's deliberations would emerge recommendations for new systems where needed to further collaboration in the many disciplines concerned.

The initial members of CSA numbered 14, most of whom had taken leading parts in the 1949 Conference. Two were from South Africa—Basil Schonland and the eminent veterinary scientist, P. J. du Toit; they were elected successively as the first and second Presidents of CSA. Three were French, namely J. Millot, director of ORSTOM's institute in Madagascar and later director of the Musée de L'Homme in Paris, J. L. Trochain, director of ORSTOM's institute in AEF, and Th. Monod of IFAN. Millot was elected vice President of CSA. Two were Portuguese, A. Mendes Correa and F. J. Cambournac who was head of WHO's regional African centre at Brazzaville. Two were Belgian, J. Lebrun of INEAC and Louis van den Berg of IRSAC. One was Rhodesian, N. P. Sellick, and three were British, namely Sir Alexander Carr-Saunders, Sir Bernard Keen, who was director of the East African Agricultural and Forestry Research Organization, and Col. H. W. Mulligan from West African medical research. I was nominated as a member but was soon pressurized to accept the post of Secretary-General after agreement had been reached with the East Africa High Commission and the Colonial Office. Thus in 1951 it fell to my lot to assist the CSA in applying Smuts' holistic approach to the scientific problems of Africa which he had advocated in 1929 (Chapter 3).

To begin with we ran the affairs of CSA from Kenya where I was already equipped with residence, office, and secretary. A Belgian biologist of wide African experience, Hans Bredo, was appointed as my deputy. The Belgian government were keen to have the organizational centre of CSA in their territory and offered accommodation at Bukavu, so we moved there in 1952. The offices occupied a former hotel on a promontary jutting into the beautiful Lake Kivu, just by the origin of the Ruzizi River, and my house looked down from a hill in Ruanda across that river. This was at a time when the Mau Mau rebellion in Kenya was drawing to a close, and there was a disturbing feeling about the future of the Belgian Congo as well as of some other African countries. In the event CSA did not enjoy a long life there for obvious political reasons.

While the scientists concerned had been working up their international system of collaboration, their opposite numbers in administration and politics had been similarly engaged. In 1945 the French and British had initiated mutual discussions and two years later the Belgians, Portuguese, South Africans, and Rhodesians joined in, with the result that in the same year as CSA was set up a 'Commission de Coopération Technique pour l'Afrique' (CCTA) was also created. This was composed of formal representatives of the member governments, and a brilliant French administrator, Paul Marc Henry, was appointed as its secretary-general. The CSA and CCTA provided a useful combination in which the scientific conclusions and recommendations could be translated into administrative action with minimum delay.

Those of us who were intimately concerned in these matters had first to understand the attitudes of the different countries and personalities with which we were to deal, for the development of Africa did not necessarily mean the same thing to a Belgian, a Frenchman, a Portuguese, a Britain, a South African, or a native of Africa itself. Attitudes were to some extent based on history, and as they appeared to an ecological observer at that time could be described as follows:

The Belgian Congo had come into existence in 1885 as the private property of Leopold II, administered by his personal servants. It was not until 1908 that it became the responsibility of the Belgian Parliament and people, who received it with some reluctance. Since then the 'Sacred mission of civilization' had been carried on with ever-increasing intensification, those in charge adopting a somewhat paternal approach, perhaps best expressed in the motto 'Dominer pour servir'. The great wealth of natural resources in the Congo had allowed economic expansion to continue rapidly throughout its short history and had enabled the Belgian system to achieve results which were far-reaching. There was little encouragement for native Africans to assume political responsibility and not much higher education, but there was very strong emphasis on technology. The closely-knit administration made it easier than under the British system to introduce improved methods in the use of land. There was a high measure of dependence on the mother country, from where the scientific organizations and institutes were controlled and to a large extent financed.

The mandated territory of Ruanda Urundi, which had a higher density of population than any other single country in Africa south of the Sahara, was administered for many purposes as an integral part of the Congo. Like other mandates in Africa the system had been heavily criticized in the United Nations and, as a counter, prominence had been given to 'la thèse Belge': rather than replying to criticisms, often ill-informed, attention was drawn to the failure of other members of the United Nations in administering their own backward peoples.

The basis of the French approach was inherited from the French Revolution with its philosophy that all men are born equal. Following this it became clear from about 1850 that dependent peoples should be assimilated as far as possible into French culture and that the Empire should be an integral part of the French Republic. Thus all the overseas peoples were represented in the Parliament of the mother country, which was the law-giving body and to which all local government systems were subordinated. Although in practice it was only locally that the assimilation of 'indigènes' had gone far, this approach was reflected, for example, in the adoption of the French language at all levels of education. In her scientific work Overseas France had less money to spend in proportion than had the Belgian Congo. But, as in the Belgian system scientific and technical progress was controlled mainly from the mother country.

Under the Portuguese system 'Colonies' were not recognized, because the single state of Portugal was divided into a number of Provinces which had similar status whether they were situated in Europe, Africa, or India. There was a more extreme form of centralization than in the French territories, and every major decision was taken in Lisbon. Three principles of political unity were prominent—spiritual assimilation, administrative assimilation, and economic solidarity. In a way somewhat similar to Overseas France the population in the three Portuguese African Provinces was composed of two juridical classes: the 'assimilados', that is citizens of Portugal, and the natives. Having attained the necessary civilized status there was no barrier to any native becoming a citizen; in fact it was claimed that the policy concerning subject peoples consisted of integrating them into Portuguese customs and ways of life without any racial discrimination. The Portuguese system had been applied in Africa for a much longer period than others. During three or four centuries the process of assimilation had been gradual. The scientific institutes and departments in Portuguese Africa depended almost wholly on the mother country.

South Africa presented strong contrasts to other African powers. The Union was the only country which had attained a high degree of civilization through the application of modern science and technology, and its population had a much higher proportion of white to black than any other. There is no need to elaborate on the thesis of apartheid: it is sufficient to point out that the South Africans were determined to settle their own problems in their own way, taking advantage of experience elsewhere, and this applied to the mandated territory of South-West Africa as well as to the Union itself.

The British approach, as adopted in the nineteenth century was expressed by David Livingstone who described Central Africa north of the Zambesi as 'An open path for British enterprise in commerce and Christianity.' Somewhat later Cecil Rhodes laid emphasis on 'Philanthropy plus 5 per cent', and for a long period it was assumed that the British people had more to give to other people than to receive from them, in religion, law, administration, commerce, and all other ways of life. The later tendency to encourage the dependent territories to run their own affairs was a logical and inevitable result. The variety in systems of direct and indirect rule adopted by the British, with numerous adaptations to local conditions, contrasted with the uniformity of the Belgian, French, and Portuguese systems. There was an emphasis on building up from local culture, illustrated by the use of the vernacular in education, coupled with the elevation of the more promising individuals by higher education, in Africa or overseas. Decentralization from the metropolis, and the integration of local people into territorial government had an influence on scientific, as on all other activities. Nevertheless, the maintenance of a measure of control in London complicated this issue, and sometimes made it a little difficult for the individual worker overseas to appreciate who was his master. As one of the consequences of the variety inherent in the British system there came into being the Federation of Rhodesia and Nyasaland.

In addition to the six member governments of CSA and CCTA were other nations responsible for areas south of the Sahara. The Spanish were applying their long experience of colonial rule to Fernanda Po and Rio Muni; the Italians were preparing Somalia for self-government; the Ethiopians and the Liberians were wholly responsible for their countries; as also were at that time the Sudanese and the Ghanians after a long period of British administration.

This summary would be incomplete without reference to the United States, not as responsible for any part of Africa, but in assisting those who were. American interests were guided by the factors of defence, humanitarianism, and financial investment. The humanitarian interest was emphasized by President Truman's Fourth Point in his inaugural speech in January 1949, as follows: 'More than half the peoples of the world are living in conditions approaching misery. Their food is inadequate. They are victims of disease. Their economic life is primitive and stagnant. Their poverty is a handicap and a threat both to them and to more prosperous areas. For the first time in history humanity possesses the knowledge and skill to relieve the suffering to these peoples.'

These words, followed by considerable action by the United States Government, particularly in arranging for American specialists and technicians to work in Africa, gave much encouragement at the time when every authority in the region was short of men, money and materials for development.

With so great a variety of external and internal influences, one might say that almost every possible system of development was being tried in some part of Africa. There was, however, one point of similarity between all: in greater or lesser degree, every country had the problem of a society composed of several races. In learning how to solve that problem history was not very helpful. The Roman system of assimilation, which created their great empire, broke down eventually when assimilation was taken too far by incorporating too many primitive peoples on the fringes. Examples of mixed racial societies breaking down in more recent times were the Ottoman Empire and Palestine. History, in fact, presented a challenge to those among us who were trying to assist in the development of Africa in the middle years of this century. In the search for cohesive influences neither religion, politics, economics, nor science proved reliable in the long run.

However, there is no need to be despondent when dreams do not come true. There was in fact much achievement in this exercise of Pan-African science. Provided the records are not lost in the confusion which has been apt to overtake newly independent countries, a considerable base was laid on which the more numerous scientists of today, both indigenous and expatriate, are able to build.

The CSA in looking at the whole field of natural and human resources recognized three stages of scientific activity although no clear lines could be drawn between them. First there is fundamental or long-term research which is designed to reveal new principles; second comes technological study in which a new principle is tested in a variety of environmental conditions in order to test how far it may be applicable to development; third comes the application and adaptation of the new principle into agriculture, fisheries, forestry, industry, medicine, education, or whatever the subject may be.

The first stage of fundamental research had been largely lacking, except in South Africa, because it depends so much on academic institutions of which few existed at that time. The scientists in government departments, with few exceptions, had been almost wholly engaged in the second and third stages, relying on principles which had been revealed by long-term studies in the very different environments of temperate latitudes. In consequence a good many mistakes had been made of which examples have been mentioned in previous chapters. It was a concern of CSA to adjust this by encouraging more long-term research on key problems of which the solution might allow economic and social development to proceed more rapidly and with fewer false steps.

At the same time, however, CSA recognized that, in pursuing the second and third stages of science and technology, a great deal of experience had been gained all over Africa. Much of this was unpublished or difficult of access. Not many scientists in Africa at that time read widely in languages other than their own, and there were not many libraries where the books and technical journals could be consulted. Therefore a second and equally important aim of CSA was to put the scientific centres and individuals in closer touch with each other.

The most effective initial way of achieving the aims of more long-term research and better communications was to convene technical Pan-African meetings on subjects which were of major importance, to publish their results, and where possible to put into operation their recommendations. Even before the formation of CSA or CCTA a few inter-African gatherings had been arranged from 1947 to 1949, notably on soils, tsetse and trypanosomiasis, rinderpest, labour, rural economy, and prehistory, the last personally organized by Louis Leakey. Thereafter, until and including 1956, more than 40 major meetings were arranged, nearly all under the auspices of CSA/CCTA. It was found convenient to arrange a few of these in Europe, but nearly all were in Africa itself, in a dozen different countries and generally at centres where the subject at issue was under special study.

The subjects selected ranged widely. On the physical bases they included maps and surveys, geology, meteorology, water, and soils. Those on plant and animal resources were on wild flora, forests, wildlife, fisheries, mechanization of agriculture, and rabies. Human subjects included medical co-operation and education, malaria, food and nutrition, social sciences, rural welfare, co-operative societies, statisics, housing, labour, education, and the treatment of offenders.

In addition there were regular annual meetings of CSA at different African centres and joint meetings or conferences of the three international organizations for study and control of the Migratory Locust, Red Locust, and Desert Locust. Meanwhile CCTA met about twice each year, tending to alternate between the European capitals of its component governments; it checked up on all these activities, stimulated others and arranged finances.

Some of the inter-African conferences resulted in setting up regular systems for consultation and joint action. For example, a committee on tsetse and trypanosomiasis held five scientific meetings during this period; another on the control of epizootic diseases brought veterinarians together. Other inter-African committees were on social sciences, on statistics, and a panel of nutritional correspondents met annually from 1944 to 1946.

As focal points to ensure continuing action in some of these subjects bureaux were located at appropriate research institutes,—one for animal diseases at Nairobi, another for tsetse and trypanosomiasis at Leopoldville (now Kinshasha). For geology an inter-African correspondent was appointed in the person of the well known geologist S. H. Haughton, based at Pretoria. He organized three regional committees for geology which, among other activities, did much to assist the pre-existing Association of African Geological Surveys in such activities as preparing a uniform geological map for the whole continent.

One of the most comprehensive and successful efforts was devoted to soil science, that subject being considered in a wide sense to cover land utilization and soil conservation; later it was expanded to include rural economy and the development of grasslands. An inter-African bureau with a full-time scientist in the person of Fred Fournier of ORSTOM was set up in Paris and produced a new technical journal. Also a Pedological Service was based at Yangambi in the Congo Belge, the headquarters of INEAC. In addition four regional committees for the conservation and utilization of the soil were set up, grouping together the countries and scientists concerned in Southern, Western, Central, and Eastern Africa. Together these had arranged no less than 12 regional scientific meetings by 1957 and the committee for southern Africa proved so effective that it survived the winds of change and continues to act as a scientific cohesive influence and, in some degree, as a policymaker for land use.

It may appear from this account that the quinquennium which I devoted to Pan-African science was concerned more with organization than with actual research, and that its connection with the evolution of ecology is remote. But that would be a wrong impression because the whole effort was, from many viewpoints, an exercise in ecology—the ecology of the workings of governments and of relations between scientific institutes and scientists themselves, as well as ecology of the physical, biological, and human environments that made up the complex of Africa south of the Sahara.

When in CSA's small secretariat I got down to examining the scientific facilities, country by country, and to assess how they had improved during some 20 years since the African Survey, it became apparent that much had changed for the better. Far more knowledge about the countries and their people existed, but there was no easy means of obtaining or classifying all the new information. Reference lists of publications were lacking even for scientific institutions. One of our early functions was to repair this gap, and we soon published and circulated a directory of the scientific institutes, organizations, and services in Africa south of the Sahara, and another of the scientific and technical libraries.

In the biological sciences there had been advances in plant and animal ecology and especially in agricultural and medical biology, but the facilities and staff for identification and description of species, many of which were still unknown to science, remained much as they had been over past years. Systematic biology and the detailed work in museums which it entailed had become an unpopular branch of science and needed a stimulus, so we published a list of systematic botanists and zoologists working in Africa. This at least helped to put specialists in touch with each other. Other booklets issued at this time were a list of scientific societies in Africa of which new ones were frequently being formed, a list of available maps, and a summary of the researches being conducted in the social sciences.

Such was the bread and butter of CSA's work. The jam was more interesting, and came mainly from numerous meetings. Some of these were quite small gatherings of a few specialists with a strictly limited objective. For example, one of primarily academic interest was a meeting in the presence of a few other ichthyologists at Nairobi between J. L. B. Smith who had described the first known specimen of a living coelacanth from the Comores Islands and J. Millot who had studied several other specimens collected later. For the most part however, meetings were on subjects which had pronounced economic or social implications and the discussions ranged widely in the uninhibited atmosphere of scientists seeking knowledge.

At this time the conservation of Africa's wild life, which was disappearing fast as a result of population pressure and the need for more land for agriculture, was attracting world-wide attention. There had already been two International Conferences on the protection of the fauna and flora of Africa. The first, held as long ago as 1929, had resulted in the International Convention of 1933 which had a big effect in stimulating some countries to establish Strict Natural Reserves, National Parks, Game Reserves, and Controlled Areas. There had been much experience and research in recent years and there was need to review the whole subject and bring things up to date. The third International Conference on this subject was therefore convened under the auspices of CSA/CCTA and IUCN at Bukavu in 1953. The Congo Belge was in some respects a leader among African countries in wildlife conservation, so after this conference many of its participants enjoyed a stimulating visit to the Parc National Albert

Fig. 25. Lake Naivasha in the Kenya Rift Valley, the site of introductions of aquatic fauna—prawns, Tilapia, black bass, and coypu. (Photo of December, 1927, see p. 24.)

under the leadership of Victor van Straelen. One of the problems which was emphasized was the difficulty of estimating wild animal populations, and of marking individual big game animals for later recognition in the wild; so we convened a specialist meeting on these topics in the following year and conducted some field trials of the methods then advocated.

Damage to agriculture and forestry caused by wildlife set other problems in ecology. The large herbivores, especially elephant, were apt to destroy crops and plantations by sheer weight and appetite. They had been accustomed to migrating over wide areas in ecological balance with the vegetation and with other animals great and small, herbivores, predators, and parasites. As agriculture and managed forests increased in area, so the range for wild animals was reduced. Over-population of elephant in some national parks, the consequent destruction of woodland, and the mass mortality of rhinoceros and some other big game which depended on the woodland, became topics for much argument. In such a situation it seemed to some that reduction of the elephant population was essential; to others such interference was against the whole principle of national parks and it was better to let nature adjust the ecological balance in her own way.

Pests of agriculture figured prominently in our work. In the past enormous losses to agriculture, both plant and animal, had been suffered by locust invasions but international action in locust control, based on intensive and widespread research, had already become one of the success stories of ecological management. However, another problem was becoming ever more acute, year by year, namely losses of grain harvests caused by African finches of the genus *Quelia*. This trouble had been prominent already for more than 25 years (see Chapter 3) and as 'agribusiness' became mechanized ever larger areas came under monoculture, and provided easy food for the birds. Therefore in 1955 CCTA/CSA convened a specialist meeting on *Quelia* birds which reviewed possible methods of control and initiated research which was much expanded later by FAO and other agencies. Nevertheless this problem is by no means solved even today, especially in connection with rice for which the area of cultivation in Africa is increasing steadily with the expansion of irrigation.

These and many other topics of importance to development became of absorbing interest. Looking back it is clear that CSA and CCTA were able to help progress in a number of ways. But soon, with the winds of change blowing ever more strongly, one country after another achieved independence, went its own way, and broke up the cohesion of Africa. The regional research institutes could no longer exercise their prescribed functions, especially when travel between adjacent countries became restricted. When it came to a golden handshake from the mother country the French system of overseas technical and scientific services with strong bases in France had certain advantages over the British, both to the newly independent country and to the cadre of experienced scientists who had worked overseas. In fact a good part of the French golden handshakes was in the form of continued staffing and servicing of technical services until such time as local African staff could take over effectively. Doctors from the French army medical service were still provided for overseas assignments, and so were scientists and technicians from ORSTOM in agriculture, engineering, and other technicalities. In the former British territories such an arrangement was not possible to nearly the same extent because the expert staff were for the most part servants of the former colony. Most of them resigned or retired prematurely to be replaced by Africans who were not always up to their jobs. In the Congo Belge the break was even more traumatic owing to the hasty withdrawal of nearly all expatriate personnel with the result that many chaotic situations arose. The military coups which were apt to replace the independent elected governments within a few years did little if anything to improve the social and political climate on which good science and development depend.

During the quinquennium under review CCTA and CSA performed functions which in some other parts of the world were regarded as the prerogative of the United Nations and its Specialist Agencies, but the UN agencies at that time had insufficient capacity to operate everywhere and, so long as the colonial powers were responsible, CCTA and CSA could work more efficiently. As colonial Africa became independent, and each new government became accepted into the UN fold, the new countries' connection with CCTA was severed. CSA and some of the collaborative systems which have been described above were in due course transferred to the Organization for African Unity (OAU), which however, was in no position to maintain the previous standards. There was an interregnum during which many inter-African activities were run down to a point of no return.

Later the UN agencies were able to create a new system. The regional office of WHO took on additional functions; a scientific office of Unesco was established at Nairobi; the regional commission of WMO had added strength; FAO increased its territorial activities; resident representatives of the United Nations Development Programme (UNDP) were appointed; and the United Nations Environmental Programme (UNEP) was set up with headquarters at Nairobi. However, it is doubtful how far the much enhanced assistance from outside was able to recreate the enthusiasm for self help and mutual assistance in science between countries and regions which was a feature of CSA. One particular loss, for unavoidable political reasons, was South Africa, which has the longest experience and greatest expertise in the African environment, but is no longer able to help, except in wholly informal ways.

Paul Marc Henry as Secretary-General of CCTA and I as Secretary-General of CSA had worked in close concert but after a few years it became clear that to have two officers of equal status, the scientist in Africa, the administrator in Europe, was not the most efficient system. Accordingly I withdrew. The CSA secretariat at Bukavu was placed under the CCTA secretariat in London for its last few years, and I spent a year in Grantchester writing a book (Worthington 1958) for CCTA and CSA about our international aspirations and achievements.

From my personal point of view, Pan-African science had been an episode of great interest and stimulation, but also on occasion of frustration. Above all it had reinforced my conviction, derived from experiences described in previous chapters, that, for the survival and wellbeing of the human species, the under-developed countries just as much as the over-developed, needed to take ecological principles much more closely into account. These principles included the conservation of natural resources, not their exploitation, nor even their preservation. Conservation, especially of renewable resources, was dynamic and constructive; exploitation was destructive; preservation was static.

9
Nature conservation (1957–1964)

Many people who have been keen hunters in their early years take enthusiastically to the conservation of nature later on. Sometimes there is a reversal of attitudes, from destroying to preserving life, but more usually, I think, the change in attitude involves a shift in focus from close involvement with day-to-day activities to a concern for the future—a concern that generations yet to come should be able to take pleasure and inspiration from nature and field sports in the same way that the passing generation has experienced.

Conservation of the wild resource has throughout history been the essence of life to the hunter, whether his interests were in large mammals, birds, fish, insects, or plants, although not all hunters have recognized this simple fact. Moreover, whether hunting as a professional or, as with most of us in the twentieth century, for recreation, success generally goes with an understanding of the habits of the quarry. In developed countries we are lucky that future provision for nature and sport has been deeply engrained through most of history. If that were not so our present landscapes of forest, field, marsh, and water, and their living contents, would be much less attractive. Modern agricultural, forestry, and urban technology has inevitably destroyed or downgraded some wildlife habitats, but a great deal remains and some is being recreated or even made anew, for example in water impoundments, and the management of reserved areas, some of which are even in large cities.

During the early years of this century, as the science of ecology was forming, the idea of 'Preservation of nature' became popular. Preservation was usually coupled with individual species, for the historical extinction of the Mauritius Dodo and more recently of the American carrier-pigeon, the increasing rarity of many plants and animals, demonstrated the precarious status of many species. Such concern was implicit in the titles of societies around that time—the Royal Society for the *Protection* of Birds (RSPB), the Society for *Preservation* of the Fauna and Flora of the Empire. The survival of disappearing species rightly persists as an important objective, but the focus on individual species has tended to change to focus on their habitat. The idea took root that if human interference is prevented, nature would look after itself and the rare species would have a better chance. There is, of course, a good deal of truth in that view, but ecological science soon showed that almost no natural habitat will remain as it was; it is continually changing, even in the absence of human activity, as a result of physical, chemical, and biological processes. A marsh will in time change to dry land as a result of accumulation of silt and the debris of vegetation; grassland, if left alone, will tend toward woodland, and will pass through a succession of associations of species of flora, each with its own fauna. Therefore, if the objective is to maintain the *status quo* in order, perhaps, to maintain the particular habitat of a rare bird, insect, or plant, management is needed—management based on ecological understanding of the changes to which the environment is subject. Thus the word *preservation*, which had a static ring to it, became replaced by *conservation*, which was interpreted as a dynamic process.

One of the early expressions of these changing ideas was 'Conservation through wise use' as Theodore Rooseveldt, who had been a great hunter, put it in 1910; but it was not until around the middle of the century that the concept of conservation became reflected in the titles of organizations. Thus the Nature Conservancy as the official organ in Great Britain was set up by Royal Charter dated 1949. The title was quickly copied for a non-governmental organization in the USA. An International Office for the Protection of Nature was set up in Brussels as early as 1928 and developed into the International Union for the Protection of Nature (IUPN) which was established at a historic meeting in the Palace of Fontainebleau in 1948. However in 1954 the word protection was replaced by conservation and the full title of IUCN had the addition of 'and Natural Resources'.

As Max Nicholson (1981) has pointed out in celebrating twenty years of the World Wildlife Fund, up till about the mid-fifties the conservation movement was in a missionary phase in which not many members of the scientific community participated wholeheartedly. The missionaries were rather few in number, men such as Julian Huxley, Frank Fraser Darling, and Peter Scott in England, Fairfield Osborn and Hal Coolidge in the USA, but they were beginning to create a public opinion.

Personally, while based in Africa for a decade, and deeply concerned about the conservation of nature and natural resources in that continent, I had been somewhat out of touch with these developments in Europe and America. Back in England in 1956 I lived with my family in the charming thatched White Cottage at Grantchester near Cambridge while writing *Science in the development of Africa*. Grantchester was soaked in scientific culture: Sir Arthur Tansley lived close by, weekly commuting to Oxford where he was Professor of Botany, and my contemporary zoologist Gregory Bateson, son of William Bateson who had 'discovered' Mendelism, had been brought up in the village. Gregory had forsaken zoology for anthropology and had spent years co-operating with and had married Margaret Mead. Andrew Huxley, the physiologist, who later became a

Nobel Prizeman and President of the Royal Society, had been tenant of the cottage until we lived there; and we were reminded of the arts by the church clock a few yards down the road, which from time to time stood still at 'ten to three' in memory of Rupert Brooke.

I began to wonder what to do when my book on Africa was finished. Many colonial servants of my acquaintance, including several Governors no older than myself, were by then living in retirement. A few picked up directorships or administrative work; one became chairman of his local Parish Council, another executive officer of a charity; some raised chickens or vegetables. What a waste of talent and experience among men and women with twenty or thirty years of overseas interests and responsibilities! I had no intention of following their example; nor, after a month or so, did there appear to be much chance to do so.

My former colleague and mentor on fishery research, Michael Graham, was at this time Director of Fishery Research to the Ministry of Agriculture, Fisheries, and Food, and was due to retire in a year or so's time. He lived near Lowestoft on a small-holding which he used for eco-farming. His laboratories contained some very bright research workers, but at that time there was no obvious successor as director, and Michael pressed me to take a post as his deputy with the directorship in mind. It was an attractive idea, to plunge back into my first love of aquatic research, all the more so since the oceans offered such wide scope and marine fishery research had been a leader in creating the science of ecology.

In the meantime I had rejoined my old college, Gonville and Caius, as a member of the Senior Combination Room, and one evening there was a knock on our Grantchester cottage door. In came Sir James Chadwick, the atomic physicist and one-time fellow of the college, who had recently been elected Master, and (Chubby) Stratton, Professor of Astronomy, who had once been my tutor. Some changes in College officers were in the offing and they had come to suggest that I return to Cambridge permanently and accept nomination as Senior Tutor. As we drank a bottle of claret on this idea, I warmed to the attractions of a suite of rooms in college, dinners in hall, interest in the students and college affairs. My visitors suggested that there would be plenty of spare time during vacations in which to continue overseas interests.

Soon after this a note arrived early in 1957 from Max Nicholson. We had met first over the Middle East when he was at the Ministry of Shipping during the war. He had proceeded to the Permanent Secretaryship in the Lord President's office and had been deeply involved in setting up the Nature Conservancy as one, though much the smallest, of the Governments' Research Councils. Its first Director-General was Cyril Diver, a great amateur naturalist who had been a clerk to the House of Commons, and in 1952 on Diver's retirement, Max had been persuaded to take the job. He should indeed have done so earlier but had succumbed to an attack of polio from which he recovered in full working order except for a limp. Max suggested a meeting.

At the conservancy's headquarters in Belgrave Square he outlined its rapid growth from nothing to an organization responsible for some fifty national nature Reserves in England, Wales, and Scotland, with several centres in each of these countries, four of them with research facilities, a growing staff of bright young ecologists in plant and animal science, and a powerful backing of Council and Committees. Max was primarily an administrator, though also an ornithologist and naturalist of international standing, and he was looking for a senior ecologist with full academic qualifications to come in as chief scientist. The job would include supervision of the Conservancy's two branches of Research and Conservation, the legal and land agency aspects being already organized by P. H. Cooper who had come to the Conservancy as Max's administrative deputy. In a sense the scientific staff in both the conservation and research branches were all research men since there were as yet no recognized procedures in management of conserved areas: techniques had to be worked out from detailed ecological study in each area. The research staff however could devote themselves wholly to their specialities, whereas the conservation staff had to be men and women of all work, and were posted to the Conservancy's regions of which there were six in England, two in Wales, and four in Scotland. One day they were perhaps exploring an area of special scientific interest in company with local naturalists, identifying unusual plants or insects; the next they were in discussion with a landlord, farmer, or gamekeeper on how best to conserve the wild plant and animal life without interfering unduly with economic or sporting values.

Of these three offers of employment the first two had the greater attractions from many points of view, but the third offered the biggest challenge and new interests in a field with which I was deeply interested but not specially familiar. So, after some further consultations, I accepted the post of Deputy Director-General (Scientific) in the British Nature Conservancy.

Grantchester was deemed too far from Belgrave Square as a place of residence. My wife needed a farm as well as a piano to exercise her talents. Our three daughters were experienced horsewomen, working or training in biological subjects (though one subsequently turned artist); and I was keen on field sports as well as nature. Our base therefore had to be rural, and search discovered an isolated farm called Colin Godmans in Sussex. On first visiting it, I met its then owner, a New Zealander, with the words 'Mr Godman, I presume?'. He looked sadly at me and intoned 'I regret to have to inform you, Sir, that the late lamented Mr Colin

Godman died early in the fifteenth century.' Indeed, the house and barn appeared little altered since those days. It was 'ripe for modernization', full of ancient features, which included a central courtyard open to the sky, only 5 m square. Colin Godmans has continued as home since that day.

Work at the Nature Conservancy started with a series of lightning tours in order to examine problems on the ground. There were meetings with staff and local personalities, including university professors and amateur naturalists, who were often potent local forces. It was indeed the amalgam of the professional and the amateur that gave the conservation movement its strength. From a strictly scientific viewpoint this was held by some to be also its weakness, because emotive overtones were apt sometimes to cloud objective analysis and synthesis of the facts. However, without the amateur naturalist, that is the person who may know his speciality as well as any professional but does not happen to have degrees in biology, the Nature Conservancy would have progressed much more slowly, especially in these early years.

There was always some prominent person to put a case for conservation. The Chairman of the Conservancy when I joined it was Sir Arthur Duncan, well known and respected among landlords and the farming and forestry communities. He was succeeded briefly by Lord Hurcomb who was recognized as spokesman on conservation in the House of Lords, and then by Lord Howick (Sir Evelyn Baring), a keen naturalist who had helped me in Africa when he was Governor of Kenya. On television there were Peter Scott and others who held the eyes and ears of millions. Max Nicholson was a power in himself, being well known throughout the senior civil service and not readily overridden in discussion. There were in fact some quite hot discussions and confrontations which can be illustrated by three examples. The first concerned Dungeness in Kent.

This promontory, which had been built up from the cross-shore movement of sand and shingle, is the corner of England nearest to the continent and as such much favoured by migrating birds. Its shingle banks and enclosed lagoons provide unique ecosystems which were under study by geographers, plant ecologists, entomologists, and ornithologists, and a bird observatory had made notable discoveries over the years. Accordingly the whole area was on the Conservancy's priority list for acquisition as a National Nature Reserve.

In 1959 it came to our knowledge that the Central Electricity Generating Board (CEGB), in their search for appropriate sites for atomic power-stations, to be located as far as possible from major human settlements and with abundant cooling water, also had Dungeness near the top of their list. Arguments against their project advanced by the Conservancy were not taken very seriously to begin with, so a public inquiry was forced and the two government agencies, the CEGB and NC, went at it hammer and tongs amidst considerable publicity. Most of us concerned thought it probable that the final decision would go in favour of the power-station being built, as indeed proved to be the case, but the CEGB found Max Nicholson as a witness so tough that never again did they put forward a proposal without full consultation with the NC in advance.

A second example, which was argued with even more publicity a few years later, was the construction of a reservoir in upper Teesdale in Yorkshire. At the source of the Tees in the Pennines is situated Moor House Nature Reserve. This is a former grousemoor of wild moorland with a shooting lodge converted into a research station which is one of the highest residential buildings in England. It is much used in ecological research by visiting scientists, especially from the University of Durham, as well as by the Nature Conservancy's own staff. Below Moor House the young river flowed down a wild and uninhabited valley before plunging over a waterfall below Cow Green. This area was much appreciated for its scenic beauty as well as for colonies of small rare plants which occurred in places where flushes of water originated from limestone and produced soils of a marked contrast to the surrounding peat. These plants were glacial relicts which had survived in this particular area since the ice age.

At this time several large industries at the river's mouth at Teeside, headed by Imperial Chemical Industries (ICI), were seriously short of water during the summer time. Their supply was pumped from the Tees which poured an abundant surplus of water into the North Sea during winter and in times of heavy precipitation, but was reduced to an inadequate flow in times of drought. The construction of a dam upstream to store flood water had therefore been advocated by them for some years.

Meanwhile W. H. Pearsall, whose wide ecological understanding has been mentioned already in Chapters 4 and 6, and who at the time was chairman of the Nature Conservancy's Scientific Advisory Committee, was advocating water conservation in flash-flood catchments, like that of the Tees, by storage reservoirs near their sources, using the natural river-bed as the pipeline, and drawing water for supply from the lower reaches. By such means the peaks and troughs of water flow could be smoothed out by storing water in flood times and releasing it during droughts. On this principle, which happily became widely accepted in later years, the site near Cow Green on the Tees appeared to be almost ideal for a dam, provided the colonies of glacial relict plants were not destroyed. Accordingly the Nature Conservancy, after balancing the needs of Tees-side industry against their designs on the area as an addition to the Moor House Nature Reserve, and having made sure that some at least of the colonies of rare plants were unlikely to be affected, was prepared to negotiate on a reservoir with a dam at Cow Green. It appeared that much less damage would be caused there to the Tees valley as a whole, including its scientific, agricultural, and amenity interests, than at alternative sites which had been suggested lower down the river's course.

However, during an ecological survey of the land which would be lost to a reservoir above Cow Green, or damaged during the construction period, a number of new colonies of the rare plants were discovered, larger and more healthy than those which would survive above the reservoir's waterline.

This led to a major confrontation in a glare of publicity, between those who wanted water for industry and those who wanted to preserve nature. In the end, as was predictable, industry won, but severe conditions were applied during construction, and subsequently for access by the public and for purposes of management. ICI, as a demonstration of its interest in conservation as well as in industrial progress, voted £100 000 for further ecological study and management of the area.

A third controversial case, which has continued through most of the century, relates to the Grey Seal (*Halichaerus grypus*). Living on fish this species has never been a friend of fishermen. It is long-lived, breeds regularly, and, apart from man, suffers practically no predation except from killer whales, which are not very numerous. The species is thus capable of rapid population increase, but it seems that this had been avoided up to this century by a substantial number being hunted, especially by fishermen. During most of the year the seals are widely dispersed but each year from November till January the adults congregate on their selected beaches or rocky shores; the cows haul out, drop and nurture their pups, and subsequently mate with the bulls who establish territories. While ashore seals are very vulnerable and their meat and skins, especially the white pelts of the very young pups, contributed to the fishermen's income.

Going back to 1914 the original Grey Seals Protection Act was passed by Parliament, largely as a result of a statement, for which there seems to have been little evidence, that the grey seal population had been so reduced by hunting that only about 500 remained in British waters. Since the species has a distribution limited to the north Atlantic and nearby seas, and had been much hunted also in the Baltic and around Norway, it appeared to be in danger of extinction, so the Act established a close season in November and December when the seals are mostly on shore. The subsequent history can be taken as a success story of conservation, or as a glaring example of where sentimentality takes over from sense, from whichever side you look at it.

By 1920, salmon fishermen at the netting stations around the Scottish coast were already complaining at the growing number of grey seals and the damage they were inflicting; and in 1927 the Secretary of State appointed Professor James Ritchie and W. L. Calderwood to investigate. They estimated the total population around Scotland, excluding the Orkneys, Shetlands, and the Farne Islands of north-west England, at between 4000 and 5000, with an annual production of 1000 pups, and they expressed their opinion that the danger of extermination no longer existed. Complaints from fishermen continued and a second Act was passed in 1932. This extended the close season, but it also provided for its suspension for periods of 12 months at a time so that control of the population could be undertaken if necessary.

After the war interest was renewed and complaints from fishermen redoubled, for the population of seals had continued to increase. The Nature Conservancy took special interest in the grey seal as the largest British wild animal, and arranged research into populations and breeding habits in the Farne Islands, West Wales, and Scotland. The government fishery scientists based at Aberdeen paid special attention also to the food of seals and to the cod-worm, a bane to fishermen, which is transmitted by seals. In 1959 the Conservancy, with the co-operation of Government, appointed the first Grey Seals and Fisheries Committee, of which I was chairman. We took a detached and objective view of the evidence and arguments for and against controlling the seal population. We appointed a seal research officer in the person of Ted Smith who is now in charge of marine mammals at Sydney Zoo. He concentrated attention on the Orkneys, where the largest assembly of grey seals around Britain was centred; and estimated the total population in 1962 around the United Kingdom and Ireland at 36 000, of which 29 500 were based on Scotland. This led to a calculation that the population had doubled in about the previous 10 to 12 years, and to a prediction that it was likely to do so again in the next 12 years unless management was applied. We concluded that in view of the undoubted damage, especially to salmon and cod fisheries, the population had already reached an intolerable level, and recommended that steps be taken to stop any further increase by an organized cull of adult females and pups using humane methods.

The meetings of this Committee were of particular interest in that its members, and those who gave evidence, included conservationists with a leaning towards the seals rather than the fishermen—men such as Frank Fraser Darling who had written a classic book after living with a colony of seals on the island of North Rona, Professor Humphrey Hewer who had studied their breeding biology and worked out their life-table, and Leo Harrison-Matthews who was Director of the London Zoo and an authority on marine mammals. After study of the facts they came to full agreement with the conclusions.

Our committee's report (1963) was accepted by the Nature

Conservancy and the Government and steps were taken to manage the population as recommended. However, they foundered owing to emotional public reaction, especially to photographs of baby seals looking pathetic, which were widely published.

Subsequent history tended to repeat itself. Much more research was undertaken. Grace Hickling, who was one of the strongest spokesmen for seals and other wildlife on the Farne Islands, became convinced of the need for controlling the population in the interests of the seals themselves, and wrote a report with Nigel Bonner, then chief seals research officer of Government, recommending culls on the Farnes which have been duly carried out. However no organized cull was undertaken in Scotland during these years while the seal population increased to between 53 000 and 55 000 as estimated by Bonner in 1970.

In 1970 came the third Grey Seals Act, as result of a Private Member's initiative with support from the Home Office. Its primary purpose was to extend conservation and management of the seal populations throughout the country, in recognition of the fact that it was necessary to maintain a 'reasonable balance of nature' between the seals and the fish populations on which they preyed. During the following years further advice was tendered by a new seals advisory committee under the late Earl of Cranbrook; this emphasized again the need for action, as also did the International Council of Exploration of the Seas (ICES), which estimated the population to have reached about 61 000 in 1973.

The Government, being convinced that action as authorized by the Act of 1970 was needed, arranged for a major cull of seals in 1978 but in the event nothing was done. Owing to widescale propaganda and the obstructions of 'Greenpeace', the operation was called off at the last moment.

In 1980 the whole question was mulled over once again at a meeting of government scientists with those from fishing industries and conservation bodies. On that occasion, speaking for the Salmon and Trout Association, which is concerned with conservation as well as with fishing, I recalled the history of the problem. A few participants pleaded for still more research before action was taken, but the consensus for proper control of the Grey Seal population was impressive. However once again the powers that be were not prepared to face an outcry, so arrangements were made for a very modest cropping of seals which will have no significant influence on the curve of population increase. In 1981, following a revised estimate of the Scottish population which indicates a still further increase by about 7 per cent in two years, the Government had to reconsider the problem once again; but sentiment again prevailed over sense.

This grey seal story is a good example of the changing balance between predator and prey, in this case man as the predator and seal as the prey. During the nineteenth century there was overkill leading to a reduced seal population; during the twentieth there has been underkill resulting in a population which has gone far beyond what is tolerable to fishermen. But the findings of ecological research have not so far been applied in practice on account of public emotions.

Such examples illustrate highlights of conservation which were important around the middle of the century, by bringing problems to notice of the public and Government. But by no means all of the Nature Conservancy's work was controversial. During this period there was steady progress in the acquisition of National Nature Reserves, a few by purchase, including the important island of Rhum, but most by rental, or by Nature Reserve Agreement between the Conservancy and landlords. Though backed by powers of compulsory acquisition the Conservancy never actually used such powers: few owners of proposed reserves were actively in opposition, and in the case of large estates, especially in Scotland, declaration as reserves had certain advantages. Usually it was not necessary to interfere with sporting rights, while the posting of a warden by the Conservancy might save the landlord a keeper at a time when the maintenance of estates was becoming ever more difficult. Public access was by no means always a condition of nature reserve agreements, though selected areas were sometimes developed for educational purposes. In some cases, where the ecosystem was brittle and liable to damage, or where rare species of interest to collectors were located, the avoidance of public interference was one of the objects.

The number of National Nature Reserves exceeded a hundred by 1964, and since then has steadily increased to 173, covering 134 000 ha by 1981. In a second category the establishment of Local Nature Reserves by county organizations was very much encouraged and their number, area, and usage has increased in proportion. Other categories are Forest Nature Reserves, most of which are established in co-operation with the Forestry Commission, and Wildfowl Reserves. The latter provided a good example of collaboration of interests which to begin with appeared to be antagonistic. The shooting around the British coasts of wildfowl, including some species of shore birds, has for several centuries provided recreation and some food for sportsmen who had no access to privately owned land and water. As the number of shooters increased and the quality of their weapons improved, the pressure on the bird populations had rendered conservation measures very desirable, but this involved restrictions which were anathema to many wildfowlers. However, the Wildfowl Association of Great Britain and Ireland (WAGBI) which at this time was starting its spectacular growth under its able director Commander John Anderton was co-operative, and over several years at the Nature Conservancy we held frequent meetings between the sporting and conserving interests. Some twenty people were present, with Max Nicholson in the chair, Peter

Scott, and Cdr Anderton as frequent spokesmen. Out of these discussions emerged reserved areas, most of them policed by members of WAGBI, where birds could roost and feed undisturbed. Shooting pressure on key areas, including some coastal Nature Reserves, was regulated, and in seasons of very severe weather shooting was voluntarily restricted or stopped. In later years members of this group took a prominent part in the work which led to the International Convention on Wetlands of 1971, and so to the establishment of internationally recognized reserved areas of water and marsh which have allowed the population of some wildfowl, especially of migratory species, to increase significantly.

To the overall picture of nature conservation in Britain Sites of Special Scientific Interest contribute a great deal. An SSSI is an area of wood, marsh, water, moor or even a field used for grazing which has some feature of particular importance, perhaps an unusual abundance of orchids, the home of some rare insect, or a pond with a good variety of fish, dragon flies, and aquatic birds. Its protection is by no means total, but before the environment may be significantly altered, particularly in a way for which planning permission is required, the Nature Conservancy has to be consulted in advance. Well over a thousand SSSIs had been designated by 1964, varying in size from a few to thousands of ha. The number increased to more than 3700 by 1981 and thereby a great deal of interest has a good chance of being saved for future generations to appreciate. Some land-owners or farmers a part of whose land is designated as an SSSI are rather proud of the fact, but others resent the restriction on what use they may make of it.

For some time the management of Nature Reserves and other conservation areas was rather haphazard because there were no plans. There was a useful guiding principle of 'If in doubt, leave it alone'; but in a developed country like Britain, where man had influenced almost every place in one way or another, the most beautiful and most interesting reserve would be likely to change its character if the particular form of management which had kept it that way were to be discontinued.

Clearly we needed management plans for every reserve so set about preparing them. In this we took advantage of the experience in forestry science which had formalized working plans designed to improve a woodland or to retain it in regular production. The Conservancy was lucky to have as Chief Conservation Officer for Scotland Joe Eggeling, an ecologist and orinthologist who had been director of forestry in Uganda and later in Tanzania; he helped in preparing guidelines for management plans. After trial and error we finished with a formula which was adaptable to every kind of ecosystem, and by 1964 we had a management plan completed and approved for most nature reserves. Although the provisions of each plan were in no way sacrosanct, because a few years' experience could lead to modification, the plan ensured continuity and it also produced a record which became of increasing interest with each year.

During these years in which the science of conservation was being moulded in Britain and other northern temperate lands, public opinion was getting the message especially through entertainment and the educational values of wildlife. As Nicholson (1981) mentioned: 'When Aubrey Buxton started his Survival series on TV in 1961, the pundits told him that it might be viable for about a dozen programmes. It has actually run for twenty years with over 350 programmes, and its own survival is still not in doubt.'

The Nature Conservancy had no remit for work overseas but in 1960, as honorary adviser on wildlife affairs to the Overseas Development Ministry, I was invited to make a journey through old haunts in East and Central Africa in order to assess how far the changes which were taking place in the build-up to independence were affecting the wild natural resources, and what assistance to conservation might be arranged by Britain (Worthington 1961). On the whole conservation of wildlife was proceeding well at that time, for the political disasters which led, for example, to the destruction of a large part of Uganda's wild fauna, had not yet taken place. Some of our British experience, such as that in preparing management plans for reserves, proved useful overseas when applied to National Parks or game reserves.

Ecologists in East Africa were somewhat divided at that time on the question of controlling the populations of elephant in several National Parks. No longer able to migrate long distances owing to agricultural development in much of the country, these great animals had multiplied in confined areas of conservation to the point of destroying the shrubs and trees in their search for food. In consequence the dominant plant associations were changing from woodland or wooded savannah to grassland, and animals which browsed rather than grazed, notably the black rhinoceros, had died from starvation by the hundred. Some of those who had studied this problem advocated control of the elephant; others considered that the phenomenon was cyclical and that nature should be left to sort it out without human interference. In the end elephant control was undertaken in several national parks by culling whole family groups with minimal disturbance to others. After independence however, there was no more need for control, for the great increase of ivory poaching decimated the elephant populations.

Another fascinating assignment was at the request of Unesco to Jamaica where several experienced biologists were pressing for the establishment of national parks and nature reserves before the wild flora and fauna, which included many unique species, succumbed to the pressures from a rapidly increasing human population. The object was to design a map of proposed conservation areas on which the government could subsequently take action if they so wished. I spent several weeks visiting all the most beautiful and interesting corners of the island, from mountain to lowland swamp and coral reef, and we finished with an inter-Caribbean conference on conservation, which dealt with nature, archaeology, and folk dance. The attempt to integrate these unlikely bedfellows was most stimulating. Assignments of this kind gave the opportunity for translating ecological principles and ideas into practical proposals. In subsequent years it was a cause of satisfaction that at least some of them turned into actions.

During the early 1960s Ethiopia was reconstructing itself after a period of tutelage under the Italians, and Emperor Haile Selassie, recognizing that there were no national parks or nature reserves in his country, had asked Unesco to initiate a plan for their creation. Accordingly, following a general assembly of IUCN in Nairobi during 1963, a small mission led by Sir Julian Huxley visited Addis Ababa. We were given facilities to travel by air and Landrover to a number of key sites, including the Awash valley, the rift valley lakes, Lake Tsana, the River Omo valley, the Simien mountains, and we managed to include the fantastic ruins of Axum. Our report and recommendations were followed by formation of a wildlife department and creation of a number of national parks and controlled areas. At our interview with the Emperor and his tame lions he remarked at one stage; 'Professor Huxley, you must recognize that my country, unlike a number of others on this continent, suffers under the disadvantage of never having been a British Colony.' Happily the developments in conservation which were initiated under his rule have survived the Ethiopian revolution of 1972 and indeed are now in process of being reinforced, for Ethiopia has not only some of the finest scenery, but also a number of unique ecosystems, including plants and animals which are in danger of extinction.

During these years the IUCN was expanding its influence for conservation of nature and natural resources on the world scale. Its secretary-general had been Jean Paul Harroy with headquarters in Brussels, but soon after he was appointed Governor of Ruanda-Urundi, the headquarters were moved in 1961 to a converted hotel in Morges, a village on the Swiss shore of Lake Geneva. IUCN was controlled by an Executive Board composed of eminent conservationists, not all of whom were professional scientists. Lord Hurcomb, who was at the time Chairman of the Nature Conservancy, was the British member on the board and, having served his term, asked me to take his place. This led to stimulating contacts with the environmental revolution as it was developing. While the board and its several commissions on national parks, endangered species,

Fig. 26. Conference of IUCN in Greece: Hans Luther, who was chairman of Finland's IPB committee, John Corner from Cambridge, and Shelagh Wakely from London.

education, law, and ecology, did excellent work, I sometimes found myself, in the early years, standing up for an objective scientific approach to problems in which propaganda also had a part to play.

In 1958 Jean Baer, the Swiss Professor of Zoology at Neuchatel University who had been President of the International Association of Zoologists, was elected President of IUCN at a memorable ceremony in the ancient theatre at Delphi. He was ably backed up by Edward Graham, an eminent ecologist who was director of the USA federal soil erosion service, and also by François Bourlière of France who later succeeded him as president. During the years 1960 to 1964, as a member of IUCN's board and later as a vice-president, I launched and was responsible for the 'African Special Project'. This was designed to follow up the third International Conference on the protection of the fauna and flora of Africa, (see Chapter 8), and its object was to make a real impression on national attitudes to conservation in one continent. We judged that it would be more effective to concentrate on Africa for a period rather than to spread the effort thinly over the whole world. The programme was designed to include first, a year's survey of the state of conservation in all of Africa south of the Sahara; second, a Pan-African conference for every country to express its views and state its needs; and third, a series of follow-up visits from outside authorities to those countries which desired them, in order to prepare detailed conservation programmes, define projects and help to get them financed.

The African Special Project clearly needed high-level backing. Unesco was behind us, but we needed particularly FAO which had up till then been reluctant to take serious interest in wildlife as a natural resource. FAO's Director-General at the time was an Indian heavily engrossed in political issues, but his deputy was my former colleague in the Middle East, Sir Norman Wright. He came under some gentle pressure, and it was agreed that a member of FAO's forestry division, Gerald Watterston, who was interested in wildlife problems, should be seconded to IUCN to conduct the initial survey. Watterston's reports gave an up-to-date picture, and we convened the Conference at Arusha in Tanzania in 1961. In addition to good representation from nearly all the countries, it was attended by high-level conservationists including Dr Grzymek, Julian Huxley, who had himself recently made a survey of African conservation for Unesco, and Peter Scott. Public relations were also in good hands with Ritchie Calder in prominence. President Nyerere sent a telegram which was widely quoted as the Arusha Declaration on Conservation —

> The survival of our wildlife is a matter of grave concern to all of us in Africa These wild creatures amid the wild places they inhabit are not only important as a source of wonder and inspiration but are an integral part of our natural resources and of our future livelihood and well-being.
>
> In accepting the trusteeship of our wildlife we solemnly declare that we will do everything in our power to make sure that our children's grandchildren will be able to enjoy this rich and precious inheritance.
>
> The conservation of wildlife and wild places calls for specialist knowledge, trained manpower and money and we look to other nations to co-operate in this important task—the success or failure of which not only affects the continent of Africa but the rest of the world as well.

Subsequent leaders of wildlife conservation and management in Africa were generally agreed that this African project of IUCN, and especially the Arusha conference, which came at a time when many countries were hoping to boost their tourist industries, was a turning point in their attitudes to wildlife as a valuable resource rather than as a hindrance to development.

One of the more tricky problems of nature conservation in Africa which the Arusha conference helped to solve was that of the world-famous Ngorongoro crater and the Serengeti National Park. For several years previously the crater and the nearby part of the Serengeti plains, which were traditionally grazed by Masai cattle, had been included in the park. The conflict of interest between the Masai and the park authority had become acute, and various plans for settling the issue had been proposed and discarded. Then in 1957 the Fauna Preservation Society, with the object of placing the issue on a more scientific footing, had sent W. H. Pearsall to examine the overall ecology of the area. Although this was his first study in tropical Africa Pearsall's report (1951) became something of a classic: it provided background for the subsequent deliberations of a special committee set up by Tanganika government in which another visitor from England, Sir Landsborough Thompson, took a leading part. This committee recommended the excision from the national park of Ngorongoro crater together with its surrounding area of highland, and also the Olduvai gorge and part of the Serengeti plain. This area was to be administered separately as a Conservation Unit in the joint interests of wildlife and the Masai.

At about the time of the conference Henry Fosbrooke was appointed as the first Conservator of the Conservation Unit, while John Owen had been appointed as Director of Tanganyka's National Parks. It was unfortunate that these two men, both good friends of mine and each with extensive experience of Africa and Africans, Owen in the Sudan and Fosbrooke in Tanganyika, did not always see eye to eye. However, each in his sphere of influence did a fine job: Owen put wildlife conservation firmly on a scientific basis by building up the Serengeti Research Institute under Hugh Lamprey, while Fosbrooke placed the Conservation Unit on a secure footing with the support of Masai chiefs and later (1972) wrote a scholarly book on *Ngorongoro crater, the eighth wonder*.

The future of the crater and Serengeti Park was a test case for harmonizing nature conservation with local human needs and when the talking at Arusha was over a number of leading figures resorted to Ngorongoro to see and discuss its problems on the spot. There were questions of forestry and grassland management, the needs and well-being of the Masai, the use of fire, population dynamics of cattle as well as of the many species of wild herbivores and predators, poaching, especially of the dwindling rhinoceros population. We toured the 100 square miles of the crater floor as countless tourists have done since, my wife and I travelling in a Landrover with Julian and Juliette Huxley and Max Nicholson.

Of predators we watched a pride of lion consuming their zebra prey, spotted hyaena emerging from and withdrawing into their underground lairs, jackal both black-backed and golden, bat-eared fox. We drove among many species of large herbivore; close to a mother rhino with her calf, a herd of eland which is the largest of the antelopes, wildebeest, hartebeest, Grant's and Thompson's gazelle, zebra, and near the surviving patch of woodland were elephant, impala, vervet monkey, and a great troupe of baboon. We observed how each species, whether grazier or browser was selective rather than indiscriminate in its feeding habits; we discussed the estimates of animal populations made recently by Michael

Grzimek who was tragically killed in an aeroplane crash while counting animals in the crater; we marvelled at the high sustained biological productivity of the crater's grassland which had continued to support this huge number of animals from time immemorial.

The lake and wetlands within the crater were specially attractive with their flocks of flamingo, greater and lesser, filtering algae from the alkaline water, and a list of other birds long enough to fill a notebook. While taking a walk at Koitoktok springs to look at hippos and the ancient game of 'Mbau' with bowls for counters carved into a flagstone probably centuries ago, Julian and Max, both leading ornithologists, suggested seeking the lammergeyer of which a pair were known to inhabit a part of the crater's inner wall.

To the uninitiated the lammergeyer or bearded vulture is a bird the size of an eagle, a rare if not disappearing species. Not much was known about its habits except that it was alleged to feed on bone marrow, though how the marrow was extracted from the bone was something of a mystery. Our African driver and guide knew where the lammergeyer had been seen so we crossed the floor of the crater and proceeded up the lower slopes of the wall where grassland was dotted with large boulders some with white tops from guano. Then the big bird, easily identified by its large wedge-shaped tail, soared into view against the wind a hundred feet or so above ground level. It was looking downward, holding a stick-like object in its talons and appeared to be taking careful aim. After aerial manoeuvre it dropped its load which narrowly missed one of the white-capped rocks. The lammergeyer swooped down to pick up and try again, but was not allowed much time because Julian and Max, in a state of excitement, were out of the vehicle and down the grassy slope to find a litter of broken bones surrounding the rock. The lammergeyer had seemingly provided its rock target with a white bull's-eye and then, having collected bones from dead animals on the crater floor, came here for bombing practice and dinner. In due course an account of this observation appeared in *The Ibis* under the joint authorship of Sir Julian Huxley and Max Nicholson.

After the Arusha conference Watterston was seconded by FAO for a further few years as Secretary-General of IUCN and the follow-up visits for the African Special Project were conducted mainly by Thane Riney. He was another leading African wildlife ecologist, and was subsequently absorbed into FAO to initiate a group of wildlife specialists.

The year 1961 will probably go down in history as a turning point in the conservation movement, leading during the next decade to the wider concept of the Environmental Revolution. Meanwhile, however, the development of natural resources, not always accompanied by 'conservation through wise use', continued unabated and will doubtless do so until human populations cease to increase and become less clamorous for an ever higher standard of living. A growing recognition of the value of wild plant and animal life to education, science, recreation, and aesthetics, do not prevent the more obvious human needs of food, fibre, shelter, and water from leading the field in development, especially in the Third World.

The end of my work with the Nature Conservancy coincided closely with the end of that organization as an independent entity under the government, administered as one of the Research Councils. In 1965 the Natural Environment Research Council was created, with the Conservancy as one, though the largest, of its parts. In 1973 this was followed by the adoption by Mr Heath's Government of the 'Rothschild Principle', and its wholesale application to science sponsored by Government.

Up to that time the usual system of government sponsorship for research conducted by the Research Councils and Universities was to make annual overall grants, after scrutiny of programmes and estimates, and then leave it to the organization to divide it up. The Rothschild principle was to replace that system by contracts for research on particular projects. The overall objective was entirely admirable, namely to close the gulf between discovery by research and its application in practice. Such a change was doubtless much needed in many cases, especially among the big organizations, but in the case of the Nature Conservancy its effect was the opposite. Max Nicholson, with the help of his lieutenants, had welded together a system where there was little if any gap between research and its application. Although the Conservancy's staff was in two streams—research and management—there were frequent transfers between them, and in several centres the same building housed both, so that the research man or woman and the regional officer responsible for management of reserves, could hardly avoid discussing mutual problems. The Rothschild Principle involved not only a great deal of extra paperwork but it split the Conservancy into two, geographically as well as administratively. The conservation and administrative part became a statutory body under the Secretary of State for the Environment (DOE) being renamed the Nature Conservancy Council (NCC), while the research side remained with NERC as a separate Institute of Terrestrial Ecology (ITE).

After the publication of a consultative document on the Rothschild Principle, many representations were made pointing out how it would bite on different scientific organizations which had been evolved to meet their particular needs with efficiency, and urging that there should therefore be discretion in its application. But government adopted the principle hook, line, and sinker. As far as nature conservation and the environmental revolution in Britain was concerned, the strong hand of government which followed the reorganization led, perhaps, to greater awareness in the corridors of power, but, while the NCC and the ITE did good work in the years which followed, something of importance to ecology and its applications was lost.

Internationally the conservation movement led by IUCN and WWF continued its upward curve, backed (though never with adequate funds) by Unesco and later by the United Nations Environmental Programme (UNEP). A step of major importance was the publication in 1980 of the World Conservation Strategy, in which IUCN/WWF, in co-operation with Unesco, FAO, UNEP, produced a document which has been widely accepted and promises well for the future.

Looking towards the end of the century, the view may appear unduly rosy to one who now lives in the green belt of Sussex among forest and farmland with a view across the Weald to the relatively unspoilt South Downs. However, the national and international aspects of the conservation of nature and natural resources are in good hands, and there is an ever-widening recognition of its importance to the welfare of the human species.

On the world scale there have certainly been positive achievements during the past twenty years: examples are the long list of National Parks and reserves, cleaner air and less polluted water in some countries. Nevertheless, while development taking proper account of ecological principles appears to be gaining ground, losses and downgrading of the environment and its ecosystems continue greatly to outweigh the gains. An outstanding example is tropical rain forest which is being devastated in the name of development in nearly all its climatic strongholds. Not only is one of the world's most highly evolved ecosystems being converted into second-rate agricultural land of low productivity, but there is loss of the forests' uses to mankind in climatic control and in products, many of which may be as yet undiscovered. As a botanical colleague wrote to me from the tropics more than a half a century ago: 'The trees depart in flames, and nought remains to clothe our ignorance'.

At the other end of the range of environments the deserts of the world suffer likewise from human acitivity, but in the opposite sense of extending rather than reducing their extent. A desert or semidesert contains ecosystems which, like the rain forest, are inhabited by specially adapted plants and animals. When used by scattered nomadic pastoralists who move from place to place according to the availability of water and grazing, the system maintains a reasonable balance. However the natural controls of human and animal populations, which include starvation, appear so ugly to the civilized view that they are ameliorated by importing food and introducing advanced agricultural technology. It is ironic that in the long run the result may be the opposite of what is intended, namely an increase rather than reduction in the total of human misery when another run of almost rainless years occurs. Fortunately it is coming to be recognized that it rarely pays off to try to convert arid land nomads who live in harmony with their brittle environment into settled people, except when the whole structure of their lives is changed by, for example, an irrigation scheme.

Turning to the highly developed—some would say over-developed—countries such as Britain, quite different changes are taking place. For example, the cleavage between the typically rural and typically urban view of nature is now less pronounced than it was twenty years ago. The amount of time devoted in both houses of parliament during passage of the Wildlife and Countryside Bill, 1981, amid all the economic and political problems which currently beset our country, is an illustration of the wide interest in conservation which now obtains.

10
The International Biological Programme (1964–1974)

In March 1959 two eminent biologists, Sir Rudolph Peters, lately retired from the chair of Biochemistry at Oxford, and Guiseppe Montalenti, Professor of Genetics at the University of Rome, together with Lloyd Berkner, the American physicist, were travelling from Cambridge to London by train late at night, after a dinner in Gonville and Caius College. Berkner had just vacated the presidency of the International Council of Scientific Unions (ICSU) to which Peters had been elected in his place, and Montalenti was about to become president of the International Union of Biological Sciences (IUBS). They were discussing the outstanding success of the International Geophysical Year (IGY) of 1957–8 and the possibility of having an international biological year, organized in a similar way. This rail journey goes down in history as the origin of the IBP, which was formally initiated five years later.

The IGY, which had been sponsored under ICSU by the International Union of Geology and Geophysics (IUGG) had a programme of research which was adapted to the capacities of many countries at different stages of development. It had focused a co-ordinated effort on certain key problems, and by so doing was stimulating a lot of subsequent work which in time led to such big advances as the theory of plate tectonics, and to eventual agreement that continental drift, which had been hotly debated since Wegener's original hypothesis was advanced in the early 1920s, was a reality. The IGY also came to cover some more spectacular activities, such as the British Antarctic Expedition to the South Pole by Sir Vivian Fuchs, which was fully motorized and equipped for geophysical study along the route. There was a considerable spin-off from the IGY of advantage to the development of resources, mainly of minerals including oil.

Just as IGY had been related to IUGG and IUGS, so would an international biological project be related to IUBS. But the biological sciences under IUBS were even more diverse than the physical under IUGG, so the objectives of a biological project would need to be limited and clearly defined. An early suggestion was that it should be concerned especially with nucleic acid, but that field was already fully stretched. Then a focus on human populations was postulated, especially those isolated in islands, valleys or on mountains before civilization had stirred them up beyond genetic recognition. A little later it was proposed that the ecological problems associated with the conservation of nature would be a good theme.

In 1960 a preparatory committee for the project was appointed by ICSU and these and other suggestions were mulled over under the chairmanship of Montalenti. Then in 1961 a definitive step was taken by IUBS at its General Assembly held in Amsterdam. Under some pressure from Russian biologists the aims were swung 'towards the betterment of mankind', with three specific areas for action, namely conservation, human genetics, and improvements in the use of natural resources. At that IUBS Assembly Conway Waddington, who was Professor of Genetics at Edinburgh University, was elected President of IUBS; after having doubts earlier about the advisability of starting an international programme at all, from then on he took a prominent part in the discussions and actions which finally launched it. Secretary to IUBS was Professor Ledyard Stebbins, the well-known American plant geneticist who was also closely involved at this early stage.

Soon the scope of the project got closer definition as 'Biological Productivity on land, in fresh waters and the sea and human adaptability thereto' and having got some way with the scope of the project its duration had to be considered. With productivity as a major theme it was obvious that the original idea of a biological year was no use. In the physical sciences it may be possible to compare conditions from place to place and time to time by organizing simultaneous observations of selected variables at a network of stations; but growth and metabolism of plants and animals involve so many factors that some quite different systems of study were needed. For example, several annual cycles are essential before it is possible to assess with reliability the productivity of an ecosystem, or even of a single species of plant or animal within that system. Thus the IB Year had to be changed to IB Programme covering a run of years.

At about this stage it was realized that the programme would have to be divided into several sections which would bring together research workers in relevant disciplines, and there would have to be a co-ordination centre in charge of a biologist of wide interests; he would have to travel and maintain contacts not only between the various sections of the IBP but also between the national organizations for carrying it out.

Waddington proceeded to look for such a person. He and I knew each other well: we had gone up to Cambridge in the same year and as post-graduate students had lived in the same house for a while. Since then we had followed each others' careers with interest, as he, having started in geological research, switched to genetics and in later years developed his particular form of philosophical biology; while I had turned to more practical affairs. We happened to breakfast together one day in the

Athenaeum Club when he asked me if I would like to become scientific director of the IBP. I was at the time deeply involved in the affairs of the Nature Conservancy and of IUCN, and was approaching the official age for retirement, but had no intention of going on the shelf. So I said 'Yes provided you can square Max Nicholson.' Nicholson was my senior as director general of the Nature Conservancy and also much interested in the possibilities of an IBP, so Waddington squared him.

Another problem was to nominate international leaders, subsequently designated as Conveners, of the seven sections into which it seemed likely that the programme would be divided. This was achieved, at a historic and argumentative planning meeting convened by Jean Baer in the headquarters of IUCN, of which he was President. These convenerships were honorary appointments, but their tenants were later assisted by paid scientific co-ordinators.

The seven sections and the conveners who presided over them were as follows:

Productivity Terrestrial (PT): F. Bourlière of France followed in 1969 by J. B. Cragg of Canada.

Productive Processes (PP): I. Málek of Czechoslovakia.

Conservation Terrestrial (CT): E. M. Nicholson of the UK.

Productivity Freshwater (Pf):V. Tonolli of Italy, followed in 1966 by A. Hasler of USA, in 1968 by G. G. Winberg of USSR, and in 1970 by L. Tonolli of Italy.

Productivity Marine (PM): R. S. Glover of UK followed in 1966 by M. J. Dunbar of Canada.

Human Adaptability (HA): J. S. Weiner of UK.

Use and Management (UM): Ed. Graham of USA, and on his untimely death, followed by G. K. Davis of USA.

The last-named section started as *Public relations and training* under L. Stebbins but it was soon decided that training and P. R. were better looked after separately by the other sections so section UM was formed.

While these manoeuvres were afoot for creating an operative organization, ICSU appointed a Special Committee to take overall responsibility for the IBP. It was composed of representatives of ICSU's scientific unions which would be involved in the programme together with individual scientists drawn from a few countries and organizations which were expected to take a prominent part. This special committee elected its own President in the person of Jean Baer who, after a five-year term, was followed in 1969 by François Bourlière. The committee also formally appointed the conveners of the seven sections and myself as scientific director.

Among the countries which came to contribute much to the programme, Japan, Great Britain, and the Scandinavian countries were early in

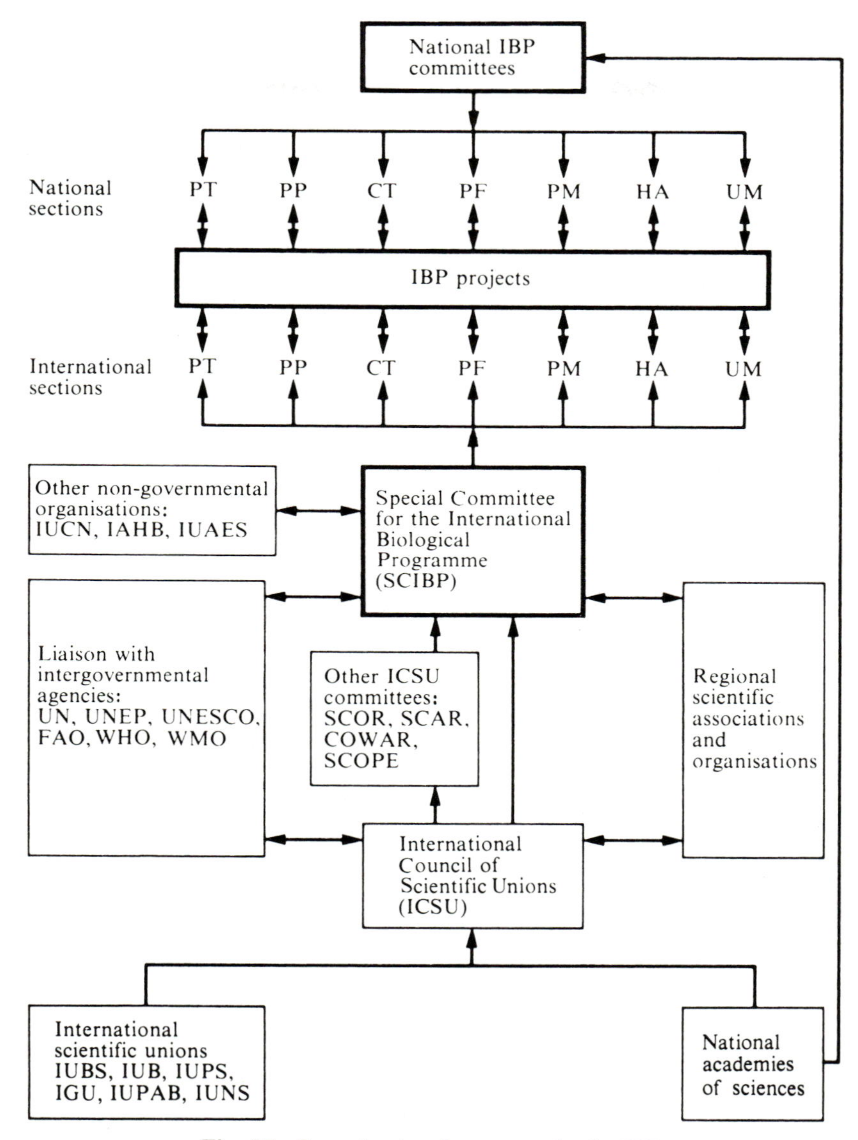

Fig. 27. Organizational structure in the IPB.

the field. The Russians were keen from an early stage but, holding themselves aloof in science as in politics, were slow to show it. Others in the Eastern bloc, notably Czechoslovakia, Yugoslavia, and Poland, showed no such inhibitions. The USA was slow to take up the challenge, but once on the move, became one of the strongest pillars of the programme.

Biologists themselves tended in the early days to follow a pro- or an anti-IBP lobby. There was to be an obvious slant towards ecology, not a subject of much consequence at that time among the scientific establishment in some countries. It did not lead to Nobel prizes, nor even except rarely to fellowships in National Academies. There was moreover sometimes apprehension that an international programme would involve competition for the limited financial resources for research. Waddington made the following comment on those early days: 'The toughest biological community into which to launch the scheme was that of the United States. The American biological world was not dominated—establishmentwise—by medical physiologists and biochemists to the same extent as the British; but the analytical school of molecular biologists and microbiological geneticists had a far higher status than in Britain and much less hesitation in asserting, in the hearing of government or the academy, that any organism bigger than *E. coli* served only to confuse the issue.'

However, after some preparation and a lot of discussions, the seed sown by Waddington and other 'founding fathers' of the IBP began to fall on fertile soil. Ecology in the early 1960s was emerging from a descriptive to an experimental science. Studies of production and population dynamics, analysis of growth, the flow of energy through ecosystems, were becoming of interest on both sides of the Atlantic, and also further afield in Japan and Australasia as well as in countries of the Eastern block. This was particularly the case in aquatic biology, both of the seas and of inland waters, perhaps because fisheries depend more directly on an understanding of natural ecological processes than do agriculture and forestry.

The countries which in the long run were likely to gain most from the kind of research on which IBP was set were those of the developing rather than of the developed world. Some of them showed a good deal of enthusiasm in the early stages, but this was difficult to maintain in the absence of funds. However, as the years went by many IBP research projects were conducted in tropical and also arctic climates, though mainly financed and staffed from countries in temperate lands, including those with former colonial responsibilities.

Finance was of course from the beginning and throughout the IBP a difficult problem. It had been hoped that, in addition to the money necessary for maintaining the small international secretariat and its world-wide functions in sectional organization and co-ordination, SCIBP would have at its disposal a substantial fund from which grants could be made for actual research projects. In the event it was never possible to establish such a fund, and this was the cause of some frustration, particularly in the work of the conservation section (CT). However, just enough was contributed from national dues, from ICSU, from contracts with Unesco, and from sale of publications, to organize the international side, the expenditure on which averaged no more than $200 000 annually. The main cost of the programme was of course on the research projects organized and paid for by the participating countries. The total number of projects proposed for inclusion in the programme exceeded 2000 but for various reasons not many more than half of them became sufficiently viable to make serious contributions.

Owing to the markedly different systems of accounting adopted by the participating countries it was never possible to assess the total cost of all the IBP research, but rough estimates put it at something in excess of $40 million per annum during the main operational phase. On any reckoning it seems that we succeeded in keeping administration to a minimum, with the maximum spent on actual research.

By July 1964 when the IBP was formally launched at the first General Assembly of SCIBP, hosted by Unesco in Paris, there had already been five years of gestation since the IUBS assembly at Amsterdam. By then many of the leading personalities knew each other well for they had worked out a comprehensive programme divided into the seven sections, and this was accepted by the Assembly. The subject of IBP was redefined as 'The Biological Basis of Productivity and Human Welfare'. Its objective was to ensure the world-wide study of (a) organic production on the land, in fresh waters, and in the seas, and the potentialities and uses of new as well as of existing natural resources, and (b) human adaptability to changing conditions. The programme did not range through the entire field of biology but was limited to the basic studies related to biological productivity and human welfare which were calculated to benefit from international collaboration, and were urgent because of the rapid rate of the changes talking place in all environments throughout the world.

There was still much to be done in preparing guidelines for the countries which would participate, and also in such matters as methodology with the object of inter-comparability of results from all parts of the world. There was need also for the training of research workers, and for dealing with such matters as the storage and retrieval of data and the publication of results. Accordingly it was arranged that the first three years should be a preparatory phase, to be followed by an operational phase extending over five years. Later another two years were added at the end in order to initiate the synthesis of results and the transfer of incompleted projects to other patronage, so that the IBP eventually had a formal life of a decade. There was however an inevitable aftermath so that a small international office was retained in London for another few years after that.

During the decade 1964–74 IBP held more than 200 international meetings in many different countries, and it published a series of 24 IBP handbooks, mainly on methods to be followed in carrying out the programme, and 25 numbers of a journal entitled *IBP News*. Some 70 volumes containing the results of symposia were also published. However, the bulk of publications which presented the results of the research in 97 countries was through the normal scientific and technical journals.

By the end of IBP the coverage of subjects was so broad that it would have been almost impossible to present the achievements to the world without a detailed analysis and synthesis. This was done during phase three of the programme, in 1973 and 1974, and the years which followed. The main synthesis was arranged in a series of 26 volumes (see reference list). This was a responsibility of SCIBP and its central office. The seven sectional conveners and scientific co-ordinators selected editors and sub-editors for each volume from those biologists who had emerged as leaders of the separate themes, and the number of contributors was considerable, in some cases exceeding fifty in a single volume, selected on their knowledge and willingness to undertake the synthesis in their particular field of work. By this means it has been possible to present the results of IBP in manageable form, occupying no more than a metre of shelf space in reference libraries all over the world.

The work done by individual countries was brought together in review papers, or, in the case of the big countries in special series of published volumes. For example, in the USSR some 43 volumes of national IBP results and synthesis have been published during and after the programme. The USA set about a series of national synthesis volumes which is not yet complete, and Japan has produced 17 volumes of synthesis in English and others in Japanese.

The most comprehensive collection of the original papers and reports was in SCIBP's central office in London. Though far from complete, this collection has been donated to the Linnean Society of London, which has one of the best-known biological libraries in Europe. Situated in Burlington House, Piccadilly, it is also handy for permanent reference.

Biological productivity and human adaptability were worked out through many specialities—biogeography and human geography; physiology and genetics of plants, animals, and mankind; variations in the physics, chemistry, and biology of the environment as affecting living organisms; the representation of ecosystems as mathematical models, and so on. Sometimes it seemed that the primary objectives were being forgotten as it became difficult to see the wood for the trees. However, there was a connecting thread for the whole in the mutual adaptation between living organism and their environment. As such the IBP presented a picture of the state of ecology at the end of the century's third quarter, and there emerged ideas on how this branch of science was likely to develop during the last quarter.

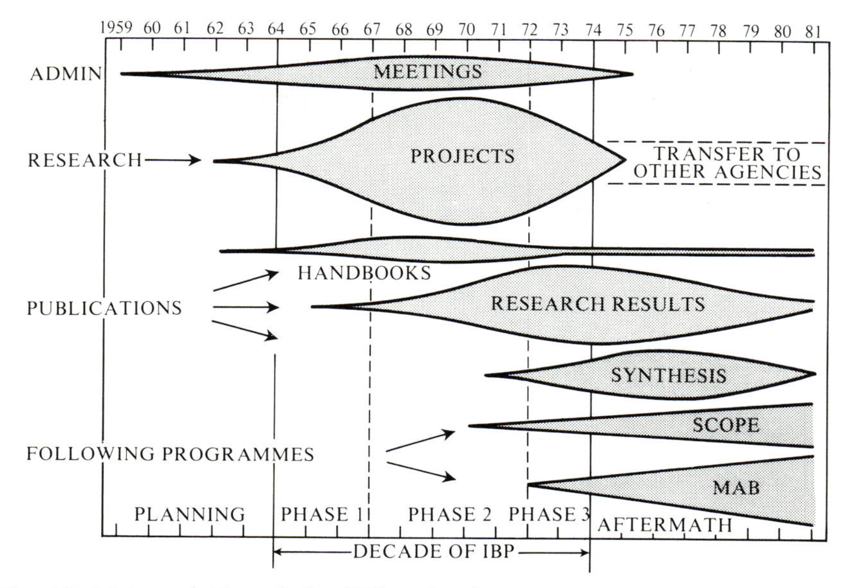

Fig. 28. Main activities of the IBP and subsequent programmes. The vertical scale is arbitrary.

The three preparatory years of Phase I were devoted to refining the programme and, in the national organizations, preparing lists of projects which were to be undertaken, arranging for field sites, finance, and staffing. Information and plans shuttled between the central and sectional offices and their national counterparts. Progress was reviewed at frequent meetings of the officers, annual meetings of SCIBP, and biennial General Assemblies which we held in several different countries in order to ensure contacts between the international centre and national organizations. Thereby the structure for the programme was built up (Fig. 27).

The scientific content of each sectional programme was elaborated, and the following summary of points which were included gives some ideas of the IBP as a whole. Since these were intentions, not all of which could be carried through to conclusion, the future tense has been retained.

Section PT (Production Terrestrial)

Investigations on the production of plant organic matter by photosynthesis and chemosynthesis at a global network of sites will cover different vegetational types.

Secondary production by animals consuming plants and each other will be studied where possible at sites where primary production is also investigated. Three main aspects concern the circulation of materials, and hence the flow of energy within ecosystems, namely (i) those related directly to man's needs, especially large herbivorous mammals, (ii) those which cause decreases or increases in primary production, for example the feeding of many invertebrates, birds, and small mammals (iii) the processes such as decomposition which occur within the soil and litter layers and are particularly important in the circulation of material.

In all cases production needs to be related to environmental factors such as solar radiation, water, micro- as well as macro-climatology.

Section PP (Production Processes)

The focus will be (i) the use of solar energy by plants in photosynthesis and (ii) the fixation of nitrogen by living organisms. The former involves fundamental physiological analysis of production per unit area by experiments conducted in agri-, sylvi-, and horti-cultural areas as well as in natural ecosystems.

Other biological processes, including phosphate metabolism and protein production by plants, will also be pursued in collaboration with other sections.

Section CT (Conservation Terrestial)

A major task will be to examine the range of ecosystems over the world and to assess how far scientifically adequate samples of all main types and variants are already protected in national parks, reserves, or research areas. Within this perspective there are questions of how to provide for many interesting species who survival cannot be ensured solely by reserved areas.

Two main processes of work are involved: (i) an inventory of major world ecosystems using vegetation formations in combination with other data, and (ii) a standard check sheet designed to record data in a rigorous and comprehensive manner about a large number of areas which are already protected or may merit protection.

Section PF (Production Freshwater)

There will be a practical slant pointing the way to means for obtaining more food by increasing productivity and more clean water by reducing productivity and pollution. To do this studies are needed on the basic factors of production and metabolism at all trophic levels, in standing and running water, whether natural, man-made, clean, or polluted.

In order to provide a base line against which the productivity and energy turnover of inland water communities can be measured it is vital to conserve and maintain examples of the different types of aquatic ecosystem, suitably distributed through the climatic zones and biogeograpical regions of the world. (*Project aqua*). In this coopeation with section CT is essential. Other contacts are with PT in transfer of material and hence flow of energy from land to water and backflow from water to land, with PP in fixation of nitrogen by algae, with PM in the transfers between inland waters and the ocean.

Section PM (Production Marine)

The basic ecological mechanisms which control the abundance, distribution, and production of marine organisms are of special importance in inshore areas where the effects of man are greatest. There are prospects for improving natural resources, for example through the cultivation of marine invertebrates, fishes, and seaweeds.

In support of the well developed international organizations in marine science, particularly in fisheries biology, there are opportunities for improving communication between marine scientists and this would be of special value to developing countries.

A central theme is the study of seasonal variation and the comparative analysis of special variation. Supporting programmes will aim at sustaining the interpretation of variation by adequate knowledge of the physiology and ecology of the most important organisms and communities.

Section HA (Human Adaptability)

The ecology of mankind in relation to diverse environments is analogous to that of plant and animal life, and vast changes affect the distribution, population density and ways of life of human communities. Some of these changes can be measured in terms of health, fitness, and genetic constitution but to do this for communities ranging from the very simple to the highly industrial requires an integrated approach and application of methods drawn from many fields.

The general aim is to elucidate physiological and genetic processes concerned in adaptation and selection in relation to climatic and other environmental factors. To achieve this a survey is needed of sample populations in conformity with a world scheme, followed by intensive multi-disciplinary regional studies based on habitat contrasts. The components would include socio-demographic and environmental assessments of the community selected for sampling, genetic constitution, medical and dental status, nutritional condition, daily and seasonal activities, assessment of physique, growth, and working capacity as index of fitness.

Section UM (Use and Management)

While some aspects of agriculture, fisheries, and harvesting of biological resources are included a strict selection is needed. Three topics are particularly appropriate in relation to other sections.

Plant gene pools are of great importance in connection with plant introductions and plant breeding, and also with evolution and taxonomy. Work will be under two headings: (1) exploration, assembly and conservation of genetic stocks, and (2) the evaluation of plant resources coupled with the biology of adaptation.

Biological control—that is the control of biota by other biota—is of growing importance owing to the many troubles which result from chemical control of pests and diseases. The programme is directed towards basic problems connected with a few selected pests of wide distribution.

The development of biological resources and nutrition, will give particular attention to foods which are rich in protein, including plants resources at present unused such as many species of leaf. Novel foods for farm animals will also be studied, as well as the differences in physiology and nutritional requirements between domesticated and wild herbivores.

Each research topic mentioned in the programme summarized above was the subject of a number of projects arranged in different parts of the world, so there was an obvious need for some uniformity in the methods used. Unfortunately in many subjects methods in current use were by no means uniform, so an important part of the phase I of IBP was devoted to the preparation and publication of a series of handbooks on methodology. There was difficulty in this because the science of ecology in most of its branches was undergoing rapid evolution. Advances were being made by the invention of new methods of research and the adaptation of others, so that any attempt to 'standardize' methods would have the effect of retarding progress. Therefore we selected leading persons as editors for the subject concerned, arranged for them to consult other specialists at their discretion, and then to describe those methods which in their view were the best or most appropriate. A foreword to each of these handbooks explained that their purpose was to advise those research workers who did not think they had better methods. Twenty-five IBP handbooks were published and the demand for some of them resulted in second and third revised editions. They continue to serve a useful purpose now, nearly a decade after the programme ended.

Phase II of the programme, entitled 'Operations', was due to commence in July 1967 by which time the sectional programmes were divided into some 80 themes and most of their scientific leaders had emerged through working groups and technical meetings. However the patterns of research did not come fully into focus until all the national projects could be tabulated, and it took time for the intentions to get around the world through institutes, universities, and individuals who were able to participate.

Ecological research can be divided into three overlapping categories: in autecology a single species is studied in order to determine its status in the environment; in synecology a number of species forming a community are studied in their relationships one to another; the third category is through the ecosystem approach in which a total biotic community together with its environment is the unit. It was in the study of ecosystems, incorporating not only the biological disciplines but also the chemistry and physics of the environment, that the IBP broke much new ground. The atmosphere, the hydrosphere, the biosphere, and the lithosphere were all in the picture.

In an entire ecosystem the number of variables which react on each other must approach infinity; so, before going into the field to collect data, it is essential to select a limited number which are likely to be of major importance as factors which influence the dynamics of the whole. Then after collecting qualitative and quantitative data, the ecosystem can be analysed and the routes of energy flow through it can be determined (Fig. 29). If there is a case for managing the ecosystem, say for the production of food or other products, the points at which management can be most effective become apparent. During this process the application of 'system analysis' to biological systems was one of the main innovations developed during the IBP. Systems analysis, which includes reducing a real system to a series of mathematical formulae and equations, has of course been a *sine qua non* for a long time in engineering, where the variables can generally be stated with some precision. Its mathematical application to biological systems, and particularly to ecosystems

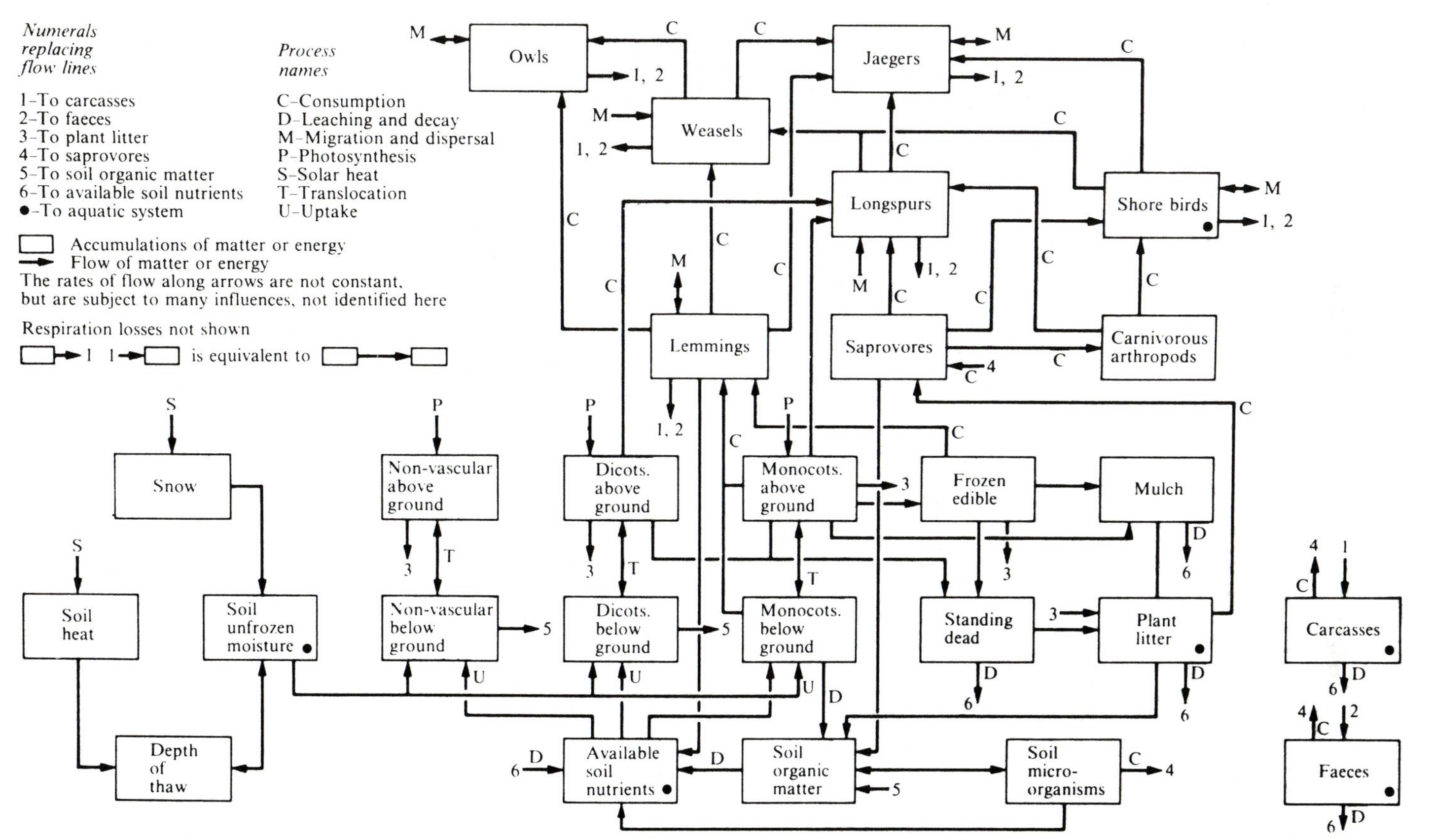

Fig. 29. Ecosystem analysis: example of the tundra near Point Barrow, Alaska. (See IBP, 1975–82, vol. 25.)

where many variables can be stated only in vague terms, had its sceptics.

However, systems analysis certainly became a factor in the IBP, more particularly in some of the projects initiated in the USA which employed large numbers of scientists and their assistants in studying the total ecology of, say, a grassland or a woodland. These came to be referred to as 'Big Biology'. In such projects, which were started rather late in the programme and continued for some years after IBP's formal ending, it was possible to apply systems analysis from the inception and design all activities to fit. Many other projects which had been initiated earlier with other techniques in mind tried later to re-orientate themselves towards dynamic systems modelling, with variable results. Since the IBP the value of systems analysis as a tool of ecology has become widely recognized, but there may sometimes be danger in the mathematical concepts dominating the biological realities.

Apart from developing new approaches to ecological research and understanding, and collecting a large amount of new knowledge, as is well documented elsewhere, what were the main successes and failures of IBP? With hindsight these have now come into focus.

A major success was undoubtedly the stimulus it gave to the broadening of ecological ideas as applied in all kinds of ways, not only to biological and human science, but to the practical affairs of mankind, in decision-making, economics, and latterly even in politics. Often the adjective 'ecological' has been grossly misused and taken far away from the objective science which it should represent, but the ecological approach to economic and social development has now become common, though by no means universal. It implies a comprehensive assessment of the consequences of development, on the environment as well as on the people, pehaps with less emphasis on cost–benefit analysis.

Another success was in IBP's continual emphasis on a fundamental approach to environmental problems, which were often being slid over in a rather superficial way during the environmental revolution. As already mentioned there was some resistance among the scientific establishment in the programme's early stages. The councils of ICSU included a strong majority of men who specialized in inanimate nature—astronomy, physics, chemistry, mathematics—and tended occasionally to resent the emergence of biology, and especially of ecology, as a main activity of their organization. There were indications at meetings that they wished to keep biology under close control. It was the fundamental approach to problems of the biosphere that overcame such criticism and led to the recognition among scientists as a whole that ecology had a major part to play in the development of science.

Another success can be claimed in the promotion of East–West co-operation. Political and language barriers did little to prevent the full and free exchange of techniques, results, and ideas, and this was largely the result of the non-governmental character of the programme. The iron curtain tended to lift during the many discussions between scientists, and IBP's leaders emerged as much from communist countries as from the democracies.

Probably the greatest success was in the way that IBP led to recognition of the importance of ecology in practical affairs. Of course it cannot claim responsibility for the environmental revolution which was gathering momentum during IBP's decade, but there is no doubt that its wide-ranging activities, coupled with the fundamental approach, had a considerable influence. Moreover IBP had a good deal to do with the creation of two major activities which succeeded it, namely ICSU's Scientific Committee for Problems of the Environment (SCOPE),

Fig. 30. IBP personalities at their last meeting in 1974 at the Royal Society, London.
Extreme back: F. Bourlière (France), President.
Back row: W. F. Blair (USA), Vice-President; M. J. Dunbar (Canada) PM Convener; J. S. Weiner (UK) HA Convener; H. Tamiya (Japan), Vice-President; R. W. J. Keay (UK), Chairman of Finance Committee.
Front row: E. M. Nicholson (UK), CT Convener; G. K. Davis (USA) UM Convener; C. H. Waddington (UK), Chairman of Planning Committee; Livia Tonolli (Italy) PF Convener: E. B. Worthington (UK), Scientific Director; Sir Otto Frankel (Australia), Vice-President.
Absent: J. B. Cragg (Canada) PT Convener; I. Malek (Czechoslovakia) Vice-President.

and Unesco's Man and Biosphere Programme (MAB), as related later.

Turning to IBP's failures and disappointments, probably the biggest was that it proved impossible to get full participation from the developing countries which stood the greatest chance of benefiting from ecological studies. There were some notable exceptions, such as projects of HA on tribal groups in a number of countries of the tropics and also on Eskimos in the Arctic; also some important studies of tropical ecosystems ranging from desert to tropical rain forest, and of some tropical waters, such as Lake George in Uganda. But the majority of projects in developing countries were undertaken by expatriate scientists and financed from the developed world. The developing countries themselves had as yet not many trained ecologists, nor the financial resources except through governments which had many more pressing demands upon them. Moreover the decade of IBP coincided fairly closely with the 'winds of change' in the former colonial world, where politics were apt to be all-absorbing, with independence achieved or just round the corner.

The turbulent times interfered with IBP's activities from time to time. For example the Middle East conflict made it very difficult for Arab to meet Jew and even led to the hijacking of the Israeli chairman of HA and his detention for some months in Syria. Nigeria had planned a substantial contribution which was greatly restricted by the Biafran war. A conference in New Delhi coincided with the outbreak of the first India–Pakistan war. The 1968 events in Czechoslovakia muzzled the active work of Ivan Malek as convener of section PP, and that country's HA chairman, the renowned physiologist Ottakar Poupa, was obliged to leave his country. Other ideological differences, the isolation of Cuba and mainland China, the restriction of movement into and out of certain countries such as East Germans to the West, South Africans to the Eastern bloc and to black Africa, limited some multinational meetings. Looking back it is perhaps surprising that IBP did not have more disappointments.

Another failure, in spite of many attempts, was to establish international data centres in which the huge quantity of information collected by IBP and other agencies, much of which would never be revealed in publications, could be stored and retrieved for future reference. Several such centres for physical data on the environment had been established for the purposes of IGY and had proved their worth; but it was difficult to provide similar facilities for biological data which, in the nature of things, are often less precise and more bulky. Something was done in this respect: a world data bank was created for conservation areas, and centres were established for certain defined subjects such as human blood genetics, but in general major international data centres for environmental biology have yet to be designed and put into being.

As the decade of IBP progressed it became clear to many that what had been initiated could not just fade away without something to follow, so in 1968 came the early moves towards the formation under ICSU of the Scientific Committee on Problems of the Environment (SCOPE) which has developed many activities during recent years. It was at the third General Assembly of IBP, held at Varna on the shore of the Black Sea, that the original fertilization of ideas took place leading to the birth of SCOPE. We had had an arduous few days on IBP and shortly before closure Bengt Lundholm of the Swedish group asked for a discussion on broad problems of the environment which in many countries were becoming acute through indiscriminate use of toxic chemicals in agriculture and control of vectors of disease, and many other actions being undertaken with inadequate study and prediction of their consequences. This was ruled as outside the province of IBP and the formal meetings were closed. However the Swedes, backed by other groups, persisted, so an informal discussion was arranged between most of those who had attended the Assembly with myself in the chair. The conclusion which we reached that some other body outside IBP was required to look into these matters was pursued later within the context of ICSU with the result that SCOPE started its active career in 1969.

At about the same time, when considering the future of ecological activities in the developing countries, it was appreciated that the advantages of a non-governmental organization might be offset by those of government sponsorship with its larger opportunities for support and for co-operation with UN agencies. Accordingly, as early as 1966, still in the preparatory phase of IBP, Michel Batisse, who was leading scientist on natural resources in Unesco, and I were discussing the possibility of following IBP with another programme organized by governments. When I put this to a meeting of IBP leaders it first had a cool reception, but soon the idea took root.

Meanwhile recognition of dangers ahead facing the world was becoming widespread. The rapid increase of human population, which had been a phenomenon of developed countries for several generations, had become obvious also in the underdeveloped world. A corresponding increase in the production of food and fibre was not keeping pace, and calculations were being published which tended to show that, with diminishing resources, it never could keep pace unless the rate of human increase slowed down. In fact a dynamic situation was arising in which conservation of natural resources was bound to assume a new dimension. By 1968, while IBP was getting into full swing, the first response of growing concern was felt by the world of international politics and Unesco convened an inter-governmental conference on the rational use of the resources of the biosphere. The delegations from governments included a fair number of IBP workers who made much of the running under the chairmanship of François Bourlière. Soon after that conference plans began to develop for the inter-governmental Man and Biosphere (MAB) programme which was launched in 1971 and has become in large measure a successor to the IBP. No time limit was set so there was no great urgency to get MAB under way, and in fact it has been rather slow in developing. Its general objective is to provide the basis for the rational use of the natural resources of the biosphere, and for the improvement of global relationships between man and his environment.

The initials of MAB are perhaps appropriate in summoning thoughts of the mythical Queen. At inter-governmental meetings, when discussion about the programme had sometimes moved from the practical to the idealistic, I turned to Shakespeare's *Romeo and Juliet* which contains a fine description of Queen MAB as the fairies' midwife, who visits men when asleep at night and delivers them of their dreams!

However, the MAB programme has already achieved a great deal more than dreams. A hundred countries are now participating, more than we had in IBP, and among its achievements to date has emerged a useful new tool for conservation and research, namely the Biosphere Reserves. The object of these is to achieve more quickly the objective that had been to the fore in the work of IUCN and section CT of IBP, namely to ensure the permanent conservation of adequate samples of every important ecosystem which has been created by evolution and adaptation. By selecting a limited number of large areas for reservation in a kind of category superior to national parks and nature reserves, and their international registration as biosphere reserves, permanent conservation is better insured and sometimes it is possible, within a single reserve, to include several different ecosystems. Each biosphere reserve is intended to contain a central wholly protected area in which evolution can continue in a truly natural state, and around this a buffer zone in which various forms of management and research can be conducted, including the monitoring of changes. Parts of the reserve can also be used for public enjoyment and education. By 1982, 58 countries had responded to the request to designate biosphere reserves, the number of which was then 214.

Biosphere reserves constitute but one branch of the MAB programme which is divided into 14 research themes relating to human interactions with major ecosystems. Although not all are yet fully operational the programme by 1980 included 900 projects which are arranged in networks covering, for example, the humid tropics, arid zones, mountains, islands, and urban systems. The relevance of MAB to the progress of ecology and to environmental problems is well explained in a short paper by Batisse (1980).

While this inter-governmental programme has advantages over IBP which have already been mentioned, it has certain disadvantages.

Notable among these is the exclusion of certain countries which, for political reasons, are not members of Unesco. Mainland China was formerly in that category, but even more important, as a leader in ecological studies, is South Africa. Another difficulty is that few governmental organizations are able to enjoy the freedom of thought and action which are often characteristic of the unofficial scientific fraternity and which were throughout a feature of IBP.

In this we are moving into the last quarter of the century, but it is fair to say that many subsequent developments which depend on ecologial understanding, and which are now coming to be regarded as international as well as national responsibilities, would not have occurred, or would have been much delayed, had it not been for the enthusiasm of non-governmental biologists who initiated and carried through the IBP. During its decade ecology emerged from description to experimentation, and it moved some way towards being a predictive science as well.

From my personal viewpoint, the post of scientific director of IBP was totally absorbing for more than a decade which spanned the years when most men have retired from full-time work and are trying to adapt to a less demanding mode of life. Moreover the programme gave the opportunity, indeed the need, to learn much new about the world, its biology, and its peoples by visiting research projects and attending meetings in more than 50 of the participating countries.

11
Fin de siècle (1975–)

The tone for the last quarter of the ecological century was set shortly before it started by the Conference on the Human Environment which was convened by the United Nations in 1972 at Stockholm. The official meetings were fraught with pomp and circumstance, but at a safe distance away in the town, there was an Environmental Forum which many participants found to offer greater entertainment. The Forum was open to all, and all had their say, sometimes sensible, sometimes silly. It was in fact a means for letting off steam by those who felt strongly about environmental issues but had no opportunity to speak at the inter-governmental meetings. The Forum even arranged a parade through the town on the slogan 'Save the Whale'.

The UN environmental conference had a very powerful effect. Not only did it force all members of the UN to think of what they were doing with their environment, but it resulted in the setting up of the United Nations Environmental Programme (UNEP) based at Nairobi. The Secretary General of UN, in an opening message, pointed out that the conference was not an isolated event 'It is a link in a chain of efforts which the UN has been making for a number of years to heighten awareness and stimulate energies within the world community in the face of great problems of our time.' This chain of effort was continued during the following decade by UN Conferences on Food, Population, Women, Habitat, Water, Desertification, and was completed in 1980 with a final one on Technology. Each of these has stimulated many studies of which a large proportion have bearings on ecology.

During this last quarter research on the biosphere continues unabated; indeed it is rapidly expanding and is particularly needed in parts of the world which have previously been studied least, between the tropics of Cancer and Capricorn. Meanwhile human ecology is coming into its own with adherents drawn from many disciplines, but where mankind is the focus of study science tends to become coloured by emotions and a truly objective approach becomes difficult to achieve.

As any science evolves from an original hypothesis, through its testing and acceptance as a theory of wide application, and then the working out of that theory in all its highways and by-ways, there comes a time when that science's exponents find themselves learning more and more about less and less. Although the scope for research in ecology is almost infinite, the overall theory of ecology has become established during the first three-quarters of this century, just as the theory of the origin of species became established a century before. Ecology is now recognized as applicable to human affairs and the time has come to translate its theory into practice.

Among the first publications to set out objectively the problem of man's influences on the terrestrial, aquatic, and atmospheric environments were what are known as the SCEP (1970) and SMIC (1971) reports from the Massachusetts Institute of Technology (MIT). Their presentation was sober with no ringing of the doomsday bell so they did not immediately make a big impression as did *Limits to growth* (Meadows, Randers, and Behrens 1972) which was stimulated by the Club of Rome. Even more widely quoted at that time was the 'Blueprint for survival' published as a number of *The Ecologist* (1972). In this document, as Lord Ashby (1977) has written in his penetrating short book on reconciling man with the environment, 'on practically every page there were assertions repugnant to the rational reader . . .' Yet Ashby also points out that the 'Blueprint' did far more to excite the public conscience than the many more restrained statements of those formative years.

We have been told many times during the last decade that 'there is only one earth'; that 'the earth is a space-ship with strictly limited resources to sustain life'; and damage to the environment caused in the processes of development have been widely publicized. However, I sometimes feel that there has been an over-emphasis on the mistakes and disasters, and an under-appreciation of the achievements which have lead to a higher quality of life. Perhaps I was out of fashion, but when invited by Barry Comoner to contribute to his symposium on the 'Careless technology' in 1968 at Airlie House, Virginia (Farvar, Taghi, and Milton 1972), I chose a success story rather than a disaster, namely the control of the River Nile. In any such example one can, of course, with hindsight, see that mistakes were made, but there was an overall benefit to mankind and to his environment. I made the suggestion that the next symposium on environmental problems might focus on the achievements of good planning rather than the mistakes, but this was received somewhat coolly.

The point of relating that episode is that the time is now overdue for bringing ecology right into development in a positive, not negative, way. I am hopeful that the last quarter of the century will see much more application of ecology in practice and there are signs that this is in fact happening. A few examples of recent events which have come my way will illustrate. They are concerned with water, which has been a recurrent theme in earlier chapters.

The perpetual passage of water through the atmosphere, biosphere, and geosphere is expressed in the simple hydrological equation:

Precipitation = percolation + Evaporation + Run-off.

Precipitation may, of course, be as snow or dew as well as rain. Percolation leads to the recharge of underground aquifers and sometimes to reappearance of water at the surface as springs. Evaporation, in addition to that from free-water surfaces, includes transpiration from plants and water-loss from animals. Run-off results in rivers, lakes, and recharge of the oceans. Thus water in one process or another flows through every ecosystem and, in addition to being one of the most essential ingredients of life, is at the heart of ecological understanding.

The assessment, development, and management of water resources for human purposes has been the special province of engineers, who have steadily enlarged and refined their technology and from time to time have turned to other disciplines for advice on problems such as of disease, pollution, or fisheries. But as development accelerated supplies of good quality water were often inadequate, and the problems to be solved became more complex. It was realized that many branches of science, fundamental as well as applied, were involved. In 1964 the Hydrological Decade was initiated by governments under the auspices of Unesco, and in the same year ICSU set up a Committee on Water Research (COWAR). For several years this committee included no biologist. Then in 1968, when the IBP was getting into its full swing, COWAR came to the attention of IUBS, and I found myself sitting as representative of biology.

During its first years COWAR had not been particularly active. One of its functions was to advise Unesco on scientific aspects of the hydrological decade, and this was done effectively in physical sciences by the International Association of Hydrological Science (IAHS) which was a component of COWAR. However, it seemed to the then President of COWAR, Professor W. Ackerman of the USA, and to several of us members, that if our committee was to serve more than as a post-box, it should do something itself of relevance to research on water resources by bringing all the scientific disciplines to bear. We looked at a number of topics which might become a special theme for attention for a year or so and settled on man-made lakes.

At this time, which was not long after Rachel Carson had stirred up feelings with her *Silent spring*, many people were becoming vocal about the effects of excessive use of modern technology on the environment. Thousands of man-made lakes, great and small, were being constructed around the world every year. Whatever their primary purpose, whether for domestic water supply, irrigation, fisheries, hydro-power, navigation, disposal of waste, every one had environmental effects, some of which were not well understood.

A lot of interest was being expressed in very large man-made lakes in developing countries, especially those in Africa where they were being constructed mainly for purposes of hydro-electric power but also for irrigation and to control floods. In newly independent countries a large artificial lake was coming almost to be a status symbol, together with an international airport and a university. Already, in connection with IBP, there had been two symposia on the environmental effects of large man-made lakes, the first in London in 1965 (Lowe-McConnell 1966) and the second in Accra in 1968 (Obeng 1969). But more information was coming to hand as those great lakes which had been created earlier, such as Kariba in Rhodesia–Zambia, Volta in Ghana, Kianji in Nigeria, Brokopondo in Surinam, were settling down, and new ones such as Cabora Bassa in Moçambique were under construction.

Controversy particularly surrounded Lake Nasser in Egypt. Some colourful but ill-considered newspaper articles had been syndicated in which the positive advantages to Egypt and the Sudan of this enormous water storage tended to be dismissed but the ancillary effects on sedimentation, evaporation, erosion, salinization of soils, fisheries, diseases, and social disruption were exaggerated. Of course there were many such problems, nearly all of which had been predicted and taken into account when the decision was taken to build the High Dam. Of course such problems would have been easier to solve had there been time to study them more thoroughly in advance; but the overall benefit of much more water for irrigation in the Sudan as well as Egypt, plus flood control in lower Egypt, is very great. Egypt did not look a gift-horse from the USSR in the mouth. In the meantime opinions expressed in the newspaper articles were accepted as fact by some who should have known better, and the High Dam at Aswan has been widely quoted as a disaster, even in ecological textbooks.

COWAR developed the subject of man-made lakes in a number of ways, by first collecting information about all those in the world of more than 100 sq. km in area which number about 600, and second by assessing the importance of small ones right down to farm-dam size, which in total have more influences on the physical, biological, and social environment than the very large lakes. COWAR's main contribution was, however, to convene the third and much the most comprehensive international symposium on the environmental effects of man-made lakes. This was held at Knoxville in 1971 with the Tennessee Valley Authority as hosts. The effects of impoundments on all aspects of the environment were thoroughly examined in general terms and specifically through selected case studies. The results fell into three categories: the physical effects, including those on climate, hydrology, sedimentation, and earth movements; biological effects including water quality, productivity, fisheries, diseases, conservation of nature; and sociological effects, including loss of land and disruption of settlements, opportunity for new agriculture and fisheries, and recreation. Following this sequence, the large volume which resulted from this symposium had three editors–Ackerman, White, and Worthington (1973). The results were further studied by a working group selected by Gilbert White, who was prominent in and later President of SCOPE, and this resulted in another publication (SCOPE 1972).

About this time a large engineering organization, the International Commission on Large Dams, selected the environmental aspects of reservoirs as a main subject for its Assembly held in Madrid. This discussion was organized by General F. J. Clarke who, as head of the United States Corps of Engineers had been roundly criticized in some quarters for seriously damaging the United States environment by putting too much of it under water. Clarke arranged for four leading engineers on one side and four environmentalists on the other to face each other on the platform, and some of the several hundred engineers in the hall were rather hoping for a kind of mental punch-up. In the event the eight protagonists, of which I was one, tended to agree much more than was expected, and the discussion was continued for an extra half day so that more of those attending could express their views.

In 1972 I followed Ackerman as President of COWAR and we set about another exercise, this time on the environmental effects of irrigation schemes in arid lands. This resulted in an international symposium comparable with that on man-made lakes; it was held at Alexandria in 1976, hosted by the Egyptian National Academy of Science. This occasion was notable for the very full coverage of the studies in Russia, especially on the great rivers including the Volga and Syr Darya which flow southwards into the Caspian and Aral Seas. Water from these rivers is stored in great reservoirs, used for irrigation, drained, re-stored, and re-used several times down the rivers' length, with consequent increase in salinity and a strict limitation of the range of crops which can be grown. The Caspian and Aral seas are steadily shrinking as a result of the high proportion of the affluent rivers' water being evaporated by the transpiration of irrigated crops. Once again the results of this symposium were reviewed by a working group under the guidance of Gilbert White. Two publications resulted, one the full record which included much original work (Worthington 1977), and a shorter document which was published by Unesco (White 1977).

In addition to salinization of soils, which is one of the most troublesome effects of irrigation in arid lands, important problems arise from water-related diseases, especially schistosomiasis; but by judicious application of ecological knowledge, involving close collaboration between engineers, agriculturalists, medical men, and biologists, maximum advantage can be taken from the beneficial effects of irrigation schemes, and the troubles can be reduced. However, it was continually emphasized at the symposium that the key factor is the attitude and actions of the individual farmers and their families. The question can be asked—are irrigation schemes made for people, or people made for irrigation schemes? And this raises a multitude of problems in human ecology.

It must be emphasized that, while in both these COWAR symposia a lot of original research was presented, the emphasis was on the practical application of the knowledge gained, as a co-operative effort between the disciplines involved. The approach was essentially ecological. These activities did not, however, receive universal approval and it came as a surprise to learn that COWAR's work was being criticized by certain elements in ICSU; not, as might be expected, for insufficient activity but for too much. Views were expressed that COWAR should revert to being a post-box for the exchange of information; so a well-known engineer, Professor Jim Dooge of Ireland, was asked to undertake an inquiry. He concluded that the COWAR purview should be enlarged to embrace more disciplines in its examination of problems in water research than those available within ICSU. He recommended that a new COWAR should be formed which would be responsible not only to ICSU but also to the Union of International Engineering Associations (UATI), which is the comparable umbrella organization for engineers. He also advocated closer cooperation as opportunity offers with the medical fraternity and with those concerned with conservation of natural areas.

Dooge's recommendation, which fitted well with my own ideas, was accepted by both ICSU and UATI; and Dooge himself took on the job of President of the new COWAR as my term of office expired. This is the first experiment at formal co-operation between the scientists of ICSU and engineers. Time will show whether it will bear fruit.

Other opportunities were offering for assisting the application of ecology. In 1969 the Commonwealth Human Ecology Council (CHEC) was formed from a group of people concerned in economics, history, religion, medicine, education, planning, as well as in ecology itself. Under the chairmanship of Sir Hugh Springer, and with the stimulating drive of Zena Daysh as secretary-general, CHEC, which has its main base in London, has extended its influence to many countries of the British Commonwealth and several others as well. It stimulates the establishment of case studies in different human communities, both urban and rural, and through conferences held in different countries. CHEC's impact has been significant also at several of the series of United Nations conferences which followed the Environmental Conference in 1972, notably at the Habitat Conference held in Vancouver in 1976.

Related to human ecology, and indeed a part of it, are environmental problems of human disease, and especially the attempts to eradicate particular diseases from large areas. The outstanding success of WHO'S project of eradicating smallpox from the entire world's population, of which success was announced in 1980, has stimulated other ambitious projects. Perhaps the largest of these currently in action is that designed to eliminate onchocerciasis (river blindness) from eight countries of West Africa. The area involved is about 700 000 square km. Onchocerciasis is caused by a parasitic worm in man and is transmitted through the bites of 'blackflies', of which the most important is well named as *Simulium damnosum*. This fly lives as a larva only in turbulent water; so the disease is limited to country which is near to rivers or streams of rapid flow. The worms in man, like those of schistosomiasis, do not kill but cause serious debility: their most unpleasant effect is partial or total blindness in up to 5 per cent of the persons infected.

The objective of the project is vast: not only to eliminate the black-fly and so the disease, but also to guide resettlement of the valley lands which have been to a considerable extent vacated owing to the disease. Since much of the area concerned is under human pressure from the over-populated Sahel region further north, the ecological problems of the onchocerciasis eradication are to some extent linked with those of the Sahel.

The main method which was to be used for eradicating the disease was application from the air of a toxic chemical, abate, to all the *Simulium damnosum* breeding sites in the entire region. The cost would be very high, some two hundred million dollars in all, because operations would need to be continued for twenty years to be reasonably sure that eradication was complete. Teams of field workers, both French-speaking and English-speaking would be organized to monitor the effect of the toxic chemicals on aquatic ecology and studies would be initiated on the agricultural and social problems of resettlement.

A large international scientific advisory committee with specialists in disease control and medical entomology was set up, and also a small watch-dog ecological panel consisting of representatives of the World Bank and UNDP as major contributors to the cost, and of WHO, FAO, and UNEP. The original members of this panel included myself as representative of UNEP and we had a fascinating time visiting the area and inspecting operations, mainly by helicopter as the only effective way of travel in that rather remote region. We held discussions in the field and at meetings in the headquarters of the World Bank in Washington and of WHO in Geneva.

Several members of the panel were from the outset unhappy about certain aspects of the project, in spite of the compelling reasons for it and the excellent work being undertaken. We felt that the whole concept was not fully in accord with ecological experience. Problems of resistance by the fly to toxic chemicals were likely—and indeed had been predicted—, and we thought that reinvasion by *Simulium* of areas cleared of it seemed more likely than was bargained for, because political reasons forbade extension of the project to cover the whole region affected by the fly and the disease.

It seemed to us that there could be other ways of controlling the disease, not yet fully explored, which might be more effective and more permanent. For example, water storage by a series of relatively small dams at chosen places down the streams would obliterate most of the turbulent rapids and concentrate the breeding sites of *Simulium* to weirs where control would be easy from the ground. Such water control would be useful morover, if not essential, for small irrigation schemes when the time came for resettlement of the valleys. Some of our comments were not wholly popular with WHO as the originators of the scheme. However, the onchocerciasis eradication project cannot but do a great deal of good to the people in that region of Africa, even though some of its original precepts may have to be modified in the light of experience.

Water-related diseases, including those carried direct in drinking water, such as typhoid and dysentery, as well as those transmitted by insects or snails, such as malaria and schistosomiasis, are liable to take on epidemic proportions when the environment is altered by development projects. Prevention of such diseases has been a joint responsibility of both engineers and doctors, but the occasions when the two have got together have not been very frequent. Accordingly the Institution of Civil Engineers in London, in co-operation with the Royal Society of Tropical Medicine and the International Association for Water Pollution Research, co-operated in a symposium held in London in December 1978.

A major theme of this gathering was derived from the changing attitudes and activities in the control of tropical diseases during this century, following successive discoveries in epidemiology, ecology, and chemistry. D. J. Bradley, director of the Ross Institute, outlined this theme in his opening address and it was followed through in contributions from a number of delegates who represented different scientific and technological disciplines.

The history of malaria illustrates these changes clearly. In 1899 Ronald Ross revealed the relationship in bird malaria between *Plasmodium*, the protozoan which causes the disease, and *Anopheles*, the mosquito which transmits it, and shortly afterwards, with Manson, he showed it also in humans. These discoveries stimulated study of mosquito ecology which showed how the insects' breeding could be prevented. Malarial control came to be largely an engineering matter—drainage, alteration of water levels, and irrigation management. A film of oil over static pools or application of 'Paris Green' killed the mosquito larvae. Around settlements, gutters were cleaned and old tins which provided breeding sites for *Anopheles* were removed.

Such methods held sway until the Second World War when the great discoveries were made of sulphanilomides, penicillin, anti-malarial drugs which were more effective than quinine, and also of DDT and other persistent synthetic insecticides. Here was the panacea for both prevention and cure of malaria; so engineering and hygiene went into eclipse. The health influences of water-supply and disposal of excreta were apt to be neglected also and this reacted on a number of diseases in addition to malaria. By 1960 the environmental aspects of health had reached a very low level.

By then it was apparent that synthetic chemistry was not a wholesale panacea after all. The vectors of disease as well as the causative organisms were adapting to their changed environment by developing resistance to the chemicals. When, on top of this there was public outcry at the damage to the environment caused by some of the toxic chemicals, the engineer was brought back into the picture.

This theme can stand a lot of working out, because, given sound ecological and medical advice, there are many ways in the design of development projects whereby the engineer, in co-operation with biologist, doctor, and social worker, can break the contact between the human being, the causative organism, and the vector.

In 1976, a few years after closure of the IBP, I was once again looking forward to years of partial leisure on my Sussex farm when a letter came from a senior person in a large firm of consulting civil engineers, Sir William Halcrow and Partners, inviting me to lunch. He produced a copy of *Middle East Science* and asked 'Did you write this?'. I knew that Halcrows were conducting engineering works in many parts of the world, but was surprised at their interest in a scientific volume written some 30 years previously.

The upshot was that I became part-time environmental adviser to Halcrows. This led to some interesting visits to unfamilliar places and to a new dimension in my thinking about the uses of ecological science.

At this time those responsible for development projects of a kind which involved large-scale engineering were starting to talk about 'Environmental Impact Statements' (EIS). In the United States, where such matters were apt to be taken more seriously than elsewhere, Federal legislation making it compulsory for an EIS to accompany any major proposal was under consideration, and was soon passed into law. Overseas projects which were to be financed in whole or in part by the World Bank were already ahead in this concept, since Dr MacNamara had made a much-quoted speech, the purport of which was that no major project submitted to the World Bank for financial aid would be considered in the absence of a statement describing and analysing its environmental effects.

One of the more dramatic moments which one must suspect helped towards MacNamara's historic statement had taken place in Uganda in 1970 overlooking the Murchison Falls. MacNamara was on one of his world tours and the Uganda government, following the success of the Owen Falls project at the outfall of the Nile from Lake Victoria, had submitted their second hydro-electric project which would harness those magnificent falls. President Obote and his ministers were present to promote the scheme, but so too was the director of Uganda's National Parks, a Muganda zoologist trained at Makerere. The director was so eloquent in pointing out the damage which a hydro scheme there would do to the scenery and to wildlife in the centre of one of the most exciting national parks in Africa that MacNamara insisted on a study of the probable environmental effects before the project was considered for financing. The project was later replaced, as described in Chapter 6. The director of parks tragically died in a motor smash soon afterwards.

Personally I do not like the term 'impacts' in the context of EIS. I prefer 'effects' because an impact implies a collision. Although it is not unknown for a development project to be set on a collision course with environmental interests, nature does not work through impacts, except on rare occasions such as volcanic eruptions. Nature normally reacts by adaptation, which is a relatively slow and often a gentle process. Moreover the term 'impact' has a negative ring, implying undesirable results, whereas some influences on the environment of a development project may have positive benefits ancillary to the primary object. The purpose of an EIS should be to draw attention to how beneficial consequences of the project can be maximized and the undesirable ones minimized.

With this reservation, I would not deny that a lot of the literature on the subject which has poured out since the United States' EIS legislation is of high value. Much of it offers guidelines, and, by listing all possible consequences of development projects of different kinds, assures that no aspect is forgotten. The prediction of environmental effects is an important branch of applied ecology (see, for example, Hefferman and Corwin (1975), Munn (1979), Clark, Bisset, and Wathern (1980)).

How does a big project in a developing country arise today? The country concerned has probably become independent in recent years, and the colonial power has gone, leaving few experienced officers who know the country and its potentialities. The new men with some exceptions are too busy running their departments to think much about innovations or new projects.

Very likely one or more missions looking at future economic possibilities have been sent by the World Bank or one of the UN specialist agencies, or perhaps by a country in the northern hemisphere under a bilateral aid scheme. The reports of such missions are available; so there is probably a list of possible projects in the recipient country worthy of closer examination, perhaps a new port here, a hydro-electric project there, a new water supply and sewerage scheme for a town, and several large-scale agricultural schemes, which will be necessary if the country is to continue to feed its rapidly growing population.

Such projects have to be arranged in a priorities list so that one or two of them can go forward as formal requests for financing under the various forms of aid. At this stage officers of UN specialist agencies, headed by the Resident Representative of UNDP, who is generally the senior UN man in the country, can help a lot in fining down or elaborating particular projects.

The next stage is for the recipient country, in consultation with the financing authority, to have feasibility studies made on one or more of the projects which have been pronounced as promising for financial aid. A feasibility study is usually an exercise involving many man-months of work by specialists in half a dozen different disciplines. The task is put out to tender and often several consulting firms of international repute are asked to submit bids. Each of them would offer detailed proposals for the study with a time scale, and of course all would be costed.

When, after perhaps a year's work by the successful tenderer, the feasibility study is completed, discussions and negotiations take place on whether or not actually to carry out the project. If the green light appears and the finance is assured, consulting engineers, who may or not be those who conducted the feasibility study, are appointed and they set to work on final designs for everything that the project needs, preparing contract papers and selecting contractors for each section of the operation, and then checking the work at all stages.

I have spelled out these various processes that are necessary before a development project is started because in some quarters criticism has been expressed of engineering firms and governmental agencies which, it is alleged, have charged blindly into major schemes without due preparation and prediction of their consequences. It cannot be denied that there have been mistakes in the past, and doubtless this still occurs on occasion. But all evolution, including the evolution of development planning, and the ecological studies leading to EIS, advances on its mistakes as well as its successes. Modern procedures provide safeguards, and many a project which has passed through feasibility study is either deferred or discarded, sometimes for lack of finance, sometimes because economic analysis shows only marginal advantages, sometimes because of predicted damage to environmental interests.

Let us look at the likely environmental consequences of two recent examples which have different objectives and are drawn from different regions of the tropical world.

In 1979 a feasibility study was conducted on a project in Tanzania of which the main object was producing rice. The study was conducted by Halcrow/ULG, an independent consultancy with Halcrow bringing in the engineering skills and ULG the expertise in agriculture and rural development. The population of Tanzania is increasing rapidly and progressively more of its people are taking to eating rice as a staple food in preference to sorghum, millet, or maize. Much of the rice has to be imported at high cost, and the government aims to save this drain on funds by increasing local production.

One of the areas favoured for large-scale rice growing is the Madabira swamp which is situated near the middle of the country, not far from the main road which runs from Dar-es-Salaam to Mbeya. The new TanZam railway is also near by. The Madabira swamp, which covers some 12 000 ha is a flood-plain of the Mwima river, which joins the Great Ruaha River.

The project includes not only diverting the Mwima River into canals to irrigate a large area of rice paddies, but also, if the economics appear favourable, constructing a dam some way upstream to convert some 8000 ha of another swamp into a shallow reservoir. This would even out the flow of the river, thereby facilitating irrigation downstream, and also would provide for a useful amount of hydro-electric power. Assessing the environmental effects of this project was an exercise in human, plant, and animal ecology, in which the changes incurred could be classified either as beneficial or undesirable. Particular forms of management or modifications to the engineering plans could be proposed in order to maximize the former and minimize the latter.

Prominent among beneficial effects would be opportunities for social and economic advancement, not only of the people concerned in the project, but also of those in the surrounding country. At present the swamp is used for grazing in the dry season and for fishing in the wet season when a good proportion of it goes under water. Its development as irrigated rice paddies with a variety of other food crops, and the establishment of new settlements, would lead to improved facilities for health, nutrition, education, and training. Such progress would be maximized if a good part of the developed land were to be allocated to smallholdings, thereby encouraging individual initiative, rather than all developed as a state farm employing labour. Fishing in the swamp would be reduced, but would be amply replaced by fishing in the proposed storage reservoir. Communications would be greatly improved.

Of the undesirable effects, disturbance to existing peasant farms around the periphery of the swamp could be turned into benefit if the farmers concerned would take up smallholdings within the project. Grazing in the swamp during dry seasons would be lost, but could be replaced by clearing adjacent land of tsetse fly and by grazing fallow areas. The worst troubles expected are those which come from large-scale monoculture. *Quelia* finches are likely to be attracted in millions to the rice fields (see p. 138). Heavy application of poisons which might be needed to control the birds, together with herbicides and insecticides to deal with other troubles would inevitably damage non-target organisms and hence the ecosystem as a whole. Accumulation of salts in the soil, which is so common in tropical irrigation unless precautions are taken by installing drains, could also become a trouble, as also could malaria and schistosomiasis, which are not particularly serious in the area at present.

These were topics which needed to be examined from the environmental viewpoint, but in addition there were opportunities for improving amenity and recreational facilities, for example by preserving as many as possible of the indigenous trees and other vegetation in and around new villages, and even demarcating certain areas as woodland or nature reserves. Big game, though present in the area in small numbers, would inevitably decrease as a result of the project, but not far away is the Great Ruaha National Park which contains good numbers and variety of species.

In short, some risk of damage to the environment was inevitable, but, when the ecology was related to studies of sociology, soils, agronomy and engineering, it seemed clear that the benefits would greatly outweigh the disadvantages. However although the feasibility study was completed early in 1980, no decision has been reached at the time of writing to proceed with this project.

The second example is concerned with two new ports desired by the government of Thailand. Exports and imports are increasing and the port facilities in that country are somewhat limited except at Bangkok. Accordingly a feasibility study was conducted during 1980–81. In the southern extension of the country towards the Malay peninsula the quantity of rubber being produced from new plantations will soon cause difficulties owing to the primitive system of loading the bales into lighters and then into merchant ships which have to anchor a considerable distance off-shore. In addition to rubber there is a steady export of tin, and vegetable oil from expanding plantations of oil palm. When to these export requirements are added the growing need for imports and for the tourist industry, the case for a deep-water port on either side of the peninsula is very strong.

The shoreline facing east to the China Sea presents a contrast to that facing west to the Gulf of Bengal. The east shore is comparatively straight, often with long sandy beaches sometimes backed by dunes, with a northward movement of shore material. However there is a gap at the town and fishery harbour of Songkhla where an estuarine channel connects the sea to a remarkable series of shallow fresh and brackish water lagoons which cover some 1040 sq. km. Songkhla is the place selected for a deep-water port.

The opposite shore of the Thai peninsula, facing west, is indented with a hinterland of hills backed by the mountain range which runs down the peninsula like a backbone. Incorporated in a multitude of inlets, bays, and small islands is a large island called Phuket, which is separated from the mainland by a narrow channel of sea. It is close to the southern end of Phuket island that the most promising site for a port had been identified.

The dangers to the environment at these two sites correspond to those differences in geography. At Songkhla the unique inland lagoons with their tidal connection to the sea already receive waste from the town and fishery harbour; additional pollution from a deep-water port could be serious. There are mangrove swamps which, as breeding grounds for fishes and crustacea, help to provide a livelihood for a large part of the population. The people are indeed numerous, with some 13 000 families living on the lake shore mainly living off the fishery. On the other side of the peninsula Phuket is also an important base for fishing industry, and a marine biological research station, which depends on unpolluted sea in its neighbourhood, is situated close to the favoured site for a port. The most vulnerable ecosystems there are associated with coral reefs which are particularly susceptible to suspended silt or pollution. Clearly there were conflicting interests at both these sites and the National Environment Board of the Thai government had prepared terms of reference for an EIS to cover each site.

In looking at such problems in a country which is developing rapidly it was clear that we should be concerned not only with 'impacts' as they appear today, but also with the future of both natural and man-made environments for, say, a couple of generations ahead. By then the human population, already considerable, may be two or three times as great with a higher standard of living. The process of urbanization will have gone much further; waste products from people, agriculture, fisheries, and industry will be greater; there will be more need for recreation and amenity, and the tourist industry is likely to be several times larger than at present. All this points to more and bigger ships, and avoidance of environmental pollution will be ever more important.

Another factor to be considered at both Songkhla and Phuket was the fishing industry. In the sea this is conducted by small vessels up to 50 tonnes of which at Songkhla up to 50 off-load their catch every day. Since the landing places are on the tidal channel the waste, added to that of the town, is swept by currents either into the lagoon where it may cause pollution, or out to sea where its break-down products add to the fishery's productivity.

A deep-water port where ocean-going ships can tie up alongside needs of course a deep-water navigation channel. At both Songkhla and Phuket this implies dredging, and this is another environmental disturbance which had to be considered. The influence of a port on the adjacent land is another matter: facilities for handling and storage of cargoes, together with eating houses and other social facilities and roadways, may need little land today but are likely to require several times as much in future.

All these considerations pointed to an overriding conclusion that the ports should be built on sites which have ample room for expansion in the sea and on land, and are as far away as practicable from places where pollution could cause damage. Only thus could options remain open to meet future needs. But in any project there are constraints which prevent the ideal from being attained; there are always limitations of finance and generally limitations in the availability of sites. However, in both these projects it was possible to reach acceptable compromise. At Songkhla the plan is to build the port outside the tidal channel so that any accidental pollution will be washed by tidal currents out to sea rather than into the lagoons. At Phuket special techniques to avoid damage to the marine ecology during construction are arranged, and at both sites there will be full provision for the disposal of waste and sewage ashore while ships are in port.

Like other development projects, these proposed ports have drawn attention to the need for areas where nature can survive in its undamaged state. From this viewpoint the lagoons at Songkhla are of high importance. Once part of the shallow sea they were cut off by sand-spits and dunes, but have retained the tidal channel which causes brackish water of varying salinity to penetrate the outer and middle lagoons. Further north is a smaller lagoon of fresh water, surrounded by marshland which grades into forest. The series shows well the evolution of aquatic environments from shallow sea through brackish to fresh water, and as a result of siltation and accumulation of vegetable debris, to land. The swamps provide a unique habitat of special value as a wintering ground and migratory staging post for northern birds, and the adjoining forest may carry the last breeding populations of several nearly extinct bird species. In addition the whole area has great variety of other fauna and vegetation, both terrestrial and aquatic.

It is satisfactory that 120 sq. km of the swamp, including the northern lagoon, have already been declared a 'non-hunting area'. It seems probable that the biological quality of a larger area is such as to warrant a higher conservation status, perhaps for registration under the International Convention for the Conservation of Wetlands.

At Phuket the conservation interest is concerned especially with the beautiful coral reefs and the areas in which the marine biological research centre has already invested much in study. This centre was established in 1970 with funds from the Thai and Danish governments and a substantial area of the near-by sea, including several islands notable for their coral reefs, was declared a 'Natural Reservation' where no fishing, mining, or disturbance other than tourism and research was allowed. Although the proposed port would be quite close by, a study of tidal currents has indicated that any polluting material which might escape from the port is very unlikely to reach the reserve.

Thus in these examples, from Africa and the Far East, (and others could be citied from the Caribbean area and elsewhere), it was possible to harmonize what appeared at first to be conflicting interests between development and conservation. There are of course other examples of proposed developments in which the ecologist might find himself in a position of strong opposition, and yet others which might be proposed by ecologists themselves which would not find favour with economists or politicians.

This book, now to be closed, has interpreted the subject of ecology somewhat broadly. It has ranged over a fair spell of time, rather widely over the surface of the earth, and into a number of specialities. But, as one grows towards maturity of years, the specialities of past experiences seem less important than their integration into a philosophy for the future.

The ecologist cannot but predict that poverty, malnutrition, and sickness will continue to present problems in vast areas of the world which are occupied today by developing countries. In the developed and overdeveloped regions he cannot but see under-employment and inflation in spite of the growing realization that 'small is beautiful'. He sees the arms race continuing apace. Conservation and development of natural resources on sound ecological principles has to await its turn, for to most people it does not yet appear as urgent as other problems.

How different the future would appear if one-tenth of the money likely to be spent on defence were to be allocated throughout the world to the study and conservation of renewable resources!

Nature has an almost infinite number of secrets yet to be revealed, and since nature is changing all the time, from century to century, day to day, and second to second, mankind is never likely to catch up by revealing them all. Meanwhile, in order to ensure that nature continues to exercise its functions in the face of changes induced by man, we must, while recognizing the limitations of our knowledge, intervene often by careful management.

Paracelsus stated this clearly in the sixteenth century: 'Nature is so subtil and so penetrating in her ways that she cannot be used except by great craft for she does not openly reveal that which may be completed within her. This completion must be accomplished by man.' Paracelsus was writing during a century of alchemy but his statement was equally applicable to the Century of Industry (Briggs 1980). Change the context a little and it rings true also of the Ecological Century.

During the first half of this century man's relations with nature were frequently expressed as a conflict. Development generally involved overpowering the forces of nature and replacing the natural environment by one which is man-made. The application of ecology is changing that, and my grandfather was right when he advised 'work with nature, my boy, not against it'. Thus my philosophy was well expressed in the words of Sir Harold Hartley at a conference in Oxford which was chaired by the Duke of Edinburgh during the early years of the environmental revolution: 'Let us achieve a double partnership, of man with man and man with nature.'

That was two decades ago. Meanwhile the environmental revolution has been running in parallel with the United Nations development decades: the first which covered the 1960s aimed at self-sufficiency in economic growth; the second during the 1970s stressed the basic needs for food, shelter, health, and education. The third which was proclaimed by the General Assembly on 5 December 1980, perhaps recognizing that both the first two fell a long way short of achieving their objectives, places emphasis on a new international economic order based on North–South interdependence. This decade's policy measures include the promotion of environmental and ecological soundness of development activities, which is an indication, perhaps, that the changing winds may be setting fair for ecology during the last quarter of the century.

The overall conclusion from this study is that ecology has an ever-growing part to play in helping to lead the way from the twentieth to the twenty-first century. To recap its history: ecology arose from the evolutionary thinking of the nineteenth century during which natural history became scientific. The first quarter of the ecological century was mainly exploratory; the second was mainly descriptive, although biological exploration was still dominant in tropical and arctic zones. The third quarter saw ecology developing into an experimental science, and, as the environmental revolution got into its stride, it became organized nationally and internationally. The last quarter is seeing the wide application of ecology to environmental and human affairs, and this gives some assurance that the twenty-first century will not become one of chaos. However, two things are necessary: one that ecology must be retained as a scientific discipline and not become a means of expressing emotions; and the other that more working ecologists must have the courage of their convictions in applying the results of their studies to practical affairs.

In the double partnership it can be claimed that ecology has already achieved much in developing the second—of man with nature. However, looking around the world of the 1980s, the first partnership—of man with man would seem yet to have far to go.

References

Ackerman, W. C., White, G. F., and Worthington, E. B. (ed.) (1973). *Man-made lakes—their problems and environmental effects*. American Geophysical Union, Washington DC.

Allen, H. B. (1946). *Rural education and welfare in the Middle East*. HMSO, London.

Ashby, H. B. (1978). *Reconciling man with the environment*. Oxford University Press.

Ball, W. L. (1944). Contribution to conference on Middle East agriculture. Department of Agriculture, Cairo.

Batisse, M. (1980). Of mammoths and men. *Unesco Courier*. May, pp. 4–8.

Beadle, L. C. (1974). *The inland waters of tropical Africa*. Longman, Harlow. (Second edition, 1981.)

Beauchamp, R. S. A. (1940). Chemistry and hydrology of lakes Tanganyika and Nyasa. *Nature, Lond*. **146**, 253–6.

Briggs, Lord (1980). Metals and imagination in the industrial revolution. *J. R. Soc. Arts* **128**, 662.

Carr-Saunders, A. (1922). The population problem: *a study in human evolution*. Oxford University Press.

Carson, Rachel (1962). *Silent spring*. Riverside Press, Cambridge, Mass.

Churchill, Winston (1908). *My African journey*. Icon, London (1964).

Clark, B. D., Bisset, R., and Wathern, P. (1980). *Environmental impact assessment*. Mansell, London.

Clements, F. E. (1916). *Plant succession: an analysis of the development of vegetation*. Carnegie Institute, Washington, DC.

Corner, E. J. H. (1981). *The Marquis—a tale of Syoran-to*. Heinemann, Asia, Singapore.

Craig, J. F. *et al.* (1979). Perch in Windermere from 1967–1977. *J. animal Ecol*. **48**, 315–25.

Drude, C. G. O. (1890). *Handbuch der planzen geographie*. Engelhorn, Stuttgart.

Ecologist, the (1972). Blue-print for survival.

Elton, Charles (1927). *Animal ecology*. Sidgwick, London.

Farvar, Taghi and Milton, J. P. (ed.) (1972). *The careless technology*. Natural History Press, New York.

Fogg, G. E. (1979). *The Freshwater Biological Association 1929–79, the first 50 years*. The Freshwater Biological Association, Ambleside.

Fosbrooke, Henry (1972). *Ngorongoro—the eighth wonder*. Andre Deutsch, London.

Freshwater Biological Association. 7th to 14th and 50th annual reports, for years 1939–46 and 1979.

Fryer, G. and Iles, T. D. (1972). *The cichlid fishes of the great lakes of Africa*. Oliver and Boyd, London.

Graham, Michael (1929). *The Victoria Nyanza and its fisheries*. Crown Agents, London.

Hailey, Lord (1939). *An African survey*. Clarendon Press, Oxford.

Hefferman and Corwin (ed.) (1975). *Environmental impact assessment*. Freeman Cooper & Co., San Francisco.

Humboldt, A. von (1845–62). *Kosmos*. Entwurf einer physischen. Weltsbeschreibung, Stuttgart.

Hurst, H. E., Phillips, P., Black, R. P., and Simaika, Y. M. (1931–50). *The Nile basin*, 8 vols. Government Press, Cairo.

Huxley, Julian (1930). *African view*. Chatto, London.

——(1961). *Conservation of wild life and 'natural' habitats in Central and East Africa*. Unesco.

IBP (1975–82). Twenty-six vols. of international synthesis. of the International Biological Programme by numerous editors and contributors. 26 vols. Cambridge University Press.

Institution of Civil Engineers (1978). Engineering and water-related diseases. *Prog. Wat. Tech.* **11**.

IUCN (1980). *World conservation strategy*. IUCN, Gland, Switzerland.

Jeffries, Sir Charles (1964). *A review of colonial research*. HMSO, London.

Keen, B. A. (1946). *Agricultural development of the Middle East*. HMSO, London.

Kipling, C. and Frost, W. E. (1970). A study of . . . pike *Esox lucius* in Windermere from 1944 to 1962. *J. animal Ecol.* **39**, 115–57.

Le Cren, E. C., Kipling, C., and McCormack, J. C. (1972). Windermere: effects of exploitation and entrophication on the salmonid community. *J. Fish. Res. Bd. Canada* 819–32.

Lowe, R. H. (1952). *Tilapia and other fisheries of Lake Nyasa*. HMSO, London.

——(ed.) (1966). *Man-made lakes*. Academic Press, London.

——(1975). *Fish communities in tropical fresh waters*. Longman, Harlow.

Macan, T. T. and Worthington, E. B. (1951). *Life in lakes and rivers*. Collins, London. (Revised 1972.)

Meadows, D. H., Meadows, D. L., Randers, J., and Behrens, N. W. (1972). *The limits to growth*. Universe, New York.

Munn, R. E. (ed.) (1979). *Environmental impact assessment*. ICSU/SCOPE Pt. 190. Wiley, Chichester.

Nature Conservancy. Annual Reports for 1950 to 1972.

——(1963). *Grey seals and fisheries*. HMSO, London.

Nicholson, Max (1970). *The environmental revolution*. Hodder and Stoughton, London.
——(1981). The first world conservation lecture. Royal Institution WWF.
Obeng, L. (ed.) (1969). *Man-made lakes and environment*. Ghana University Press.
Pearsall, W. H. (1957). Report on an ecological survey of the Serengeti national park. *Oryx* **4**, 71–136.
Ricardo, C. K. (1938). *Fish and fisheries of Lake Rukwa and Bangweulu region*. Crown Agents, London.
Ricardo, Bertram C. K., Borley, H. J. H., and Trewarvas, E. (1942). *Fish and Fisheries of Lake Nyasa*. Crown Agents, London.
Russell, E. J. (1927). *Soil conditions and plant growth*. Longmans, London.
Salisbury, E. J. (1926). Geographical distribution of plants in relation to climatic factors. *Geographical J.* **67**, 312–15.
SCEP (1970). *Man's impact on the global environment*. MIT, Cambridge, Mass.
SCOPE (1972). Man-made lakes as modified ecosystems. ICSU/SCOPE 2. ICSU, Paris.
SMIC (1971). *Man's impact on climate*. MIT, Cambridge, Mass.
Stebbing, E. P. (1935). The encroaching Sahara. *Geographical J.* **85**, 506–204.
Tansley, A. G. (1923). *Practical plant ecology*. Allen, London.
Thomas, H. B. and Scott, R. (1935). *Uganda*. Clarendon Press, Oxford.
Trapnell, C. G. and Clothier, J. N. (1937). *The soils, vegetation, and agricultural systems on North-Western Rhodesia*. Report of Ecological Survey, N. Rh. Government Press, Salisbury, Rhodesia.
Warming, E. (1909). *Oecology of plants* (English edition). Clarendon Press, Oxford.
White, G. F. (ed.) (1977). Environmental effects of arid land irrigation in developing countries. *Unesco/MAB Tech. Notes* 8.
Worthington, E. B. (1929). *Fishing survey of lakes Albert and Kioga*. Crown Agents, London.
——(1932). *Fisheries of Uganda*. Crown Agents, London.
——(1939). *Science in Africa*. Clarendon Press, Oxford.
——(1946*a*). *Middle East science*. HMSO, London.
——(1946*b*). *A development plan for Uganda*. Uganda Government Press, Entebbe.
——(1950). *An experiment with populations of fish in Windermere, 1939–48* PZS, **120**, pp. 113–49.
——(1952). A survey of research and scientific services in East Africa, 1947–56. East Africa High Commission No. 6. Nairobi.

——(1958). *Science in the development of Africa.* Commission de Cooperation Technique pour' l'Afrique, London.

——(1961). *The wild resources of East and Central Africa.* HMSO, London.

——(ed.) (1977). *Arid land irrigation in developing countries: environmental problems and effects.* Pergamon, Oxford.

Worthington, S. and Worthington, E. B. (1933). *Inland waters of Africa.* Macmillan, London.

Zahlan, A. B. (ed.) (1978). *Technology transfer and change in the Arab world.* Pergamon Press, Oxford.

Index

Most places are included, but countries are omitted. Names of people are as at the time concerned, with subsequent titles etc. following in brackets.